FOREWORD

Many OECD countries have provided financial incentives for farmers to keep agricultural land out of crop production or shift it to alternative uses as part of their agricultural policy. As a result, substantial areas of farmland have been diverted from crop production, with important consequences for land use and the environment.

Land diversion schemes have been implemented to reduce the supply of agricultural commodities or to conserve the environment, and in many cases pursue both objectives. The environmental effects of these schemes depend on a number of factors, including the length of the diversion period, the type of land taken out of crop production, the rules governing the treatment of idled land, and the possibilities for alternative land use. Other factors, such as the local and regional environmental situation, the farming systems used, and the overall level of agricultural support, are also important.

In practice, the ways in which the different factors interact and influence the environment are not well understood. By analysing concrete policy experiences, this study sheds light on the environmental implications of land diversion programmes, and examines the usefulness of these programmes as a means of moving towards sustainable agriculture and improving the environment. On a broader scale, the study helps clarify the complex linkages between agricultural policies and the environment.

The study discusses the experiences that have been made with land diversion schemes in five OECD countries: Canada, the European Union, Japan, Switzerland, and the United States. The study describes first, for each land diversion scheme, the policy provisions that determine which land is withdrawn from crop production, for how long it is taken out of production, and what is to be done with this land. It then examines the changes in land use, input use and farming practices that have occurred as a result of the scheme. Finally, empirical evidence of the impacts of the land diversion scheme on soils, water, air, biodiversity, wildlife habitat and landscape is discussed. In addition, the study explores the implications of land diversion schemes for the supply of agricultural commodities, their budgetary costs, and their relationship with other agricultural support policies.

This study is the result of work carried out by the Joint Working Party of the Committee for Agriculture and the Environment Policy Committee (JWP). It is part of the larger effort to explore the linkages between agricultural policies and the environment, and to identify the environmental effects of agricultural policy reform. The two parent committees approved the report in 1996 and agreed that it be derestricted under the responsibility of the Secretary-General of the OECD.

TABLE OF CONTENTS

SUMMARY AND CONCLUSIONS

The land diversion programmes discussed in this document pursue the objective of supply control or environmental conservation, and in most cases a combination of both. They differ with respect to the length of the set-aside period, the type of land taken out of production, the rules governing the treatment of idled land, and the possibilities for alternative land use. These differences reflect the economic, environmental and policy situations that prevailed in the countries at the time when the programmes were implemented. Yet the experiences made with these programmes also reveal similarities as to the possibilities and limitations of land diversion policies as a means of achieving a sustainable agriculture and safeguarding the environment.

Many of the programmes, especially those of the European Union and Switzerland, are of relatively recent origin, and only preliminary information on their environmental effects is available. Most of the programmes have gone through several modifications since they were first implemented. There appears to be a convergence in the way the schemes have been adapted in different countries to the growing importance of environmental concerns and changes in world market conditions. This process of adaptation continues and some countries are about to enter a new phase, which is marked by the introduction of an increasing number of conservation measures in the farm sector. These measures, which can substantially vary in the way they influence land use and farming practices, are intended to address a wider range of positive and negative effects of agriculture on the environment. The future role of land diversion schemes for environmental conservation will critically depend on these policy developments.

The three most important policy criteria influencing the environmental effects of land diversion schemes are the ones that determine which land is taken out of production; for how long it is taken out of production; and what is to be done with this land. In most cases studied, the length of the set-aside period is an indication as to whether land diversion is primarily a supply control or an environmental conservation measure. Annual set-aside and multiannual set-aside of up to five years ("short-term" set-aside) are mainly aimed at supply control, whereas "long-term" set-aside of 10 years or more has chiefly environmental objectives. As a consequence, the environmental goals of short-term set-aside are usually different from those of long-term land diversion schemes.

Land diversion schemes have taken substantial areas of land out of production or shifted to alternative uses. In Canada, around 520 000 hectares of cropland have been placed under a permanent cover. In the European Union, some 7.2 million hectares of land were diverted under short-term set-aside schemes in 1995, and around 930 000 hectares have been signed up under the forestry scheme.[1] In the United States, 14.6 million hectares of land were enrolled in the *Conservation Reserve Program* in 1995, and 5.6 million hectares were idled under annual programmes. Japanese farmers diverted 660 000 hectares of paddy fields from rice production in 1995, and Swiss farmers placed 57 000 hectares of land into land diversion schemes in 1994.

SHORT-TERM SET-ASIDE

Short-term set-aside can be operated in a rotational or a non-rotational way. Rotational set-aside is a supply control tool; its management at the farm-level is largely governed by agronomic considerations. However, some environmental goals, such as improvements in soil structure and nutrient balance, can also be pursued. The land parcels are idled for one growing season and then returned to production. In successive years, a large part of a farm's land base is taken out of production irrespective of whether individual parcels are in need of environmental improvements. In many cases, a risk of environmental degradation is created on the idled land. The risks depend on the climate, the soil type and the farming system, and include wind erosion during the dry season, water erosion on steep slopes during the winter, nutrient leaching on soils that are saturated with nutrients or have a low retention capacity, invasion of weeds and pests and, in the case of rice paddy fields, deterioration of the hydrological functions of the land.

The environmental conditions attached to annual set-aside are to a large extent motivated by the need to reduce these risks. They typically involve the establishment of a plant cover during the idling period and restrictions on fertiliser and chemicals use on diverted land. Experience with annual set-aside in the European Union, Japan and Switzerland indicates that negative impacts on the environment can normally be prevented if appropriate management rules are established and observed by farmers.

In recent years, attempts have been made in some countries to not only prevent environmental degradation but also to procure specific environmental improvements on one-year set-aside. These efforts have included use of carefully composed seed mixtures for the vegetative cover to enrich the soil with nutrients and organic matter, improve the soil structure and reintegrate local plant species into the rotation. The judicious selection of grass species has also been used to generate nesting habitat and feed supply for birds. Provided that the nesting habitat is created each year during the breeding season, improvements in birdlife can be of a longer term nature even though the idled plots may be rotated annually. Set-aside after a cereal crop can provide an over-winter cover of stubble which, combined with a thin layer of natural vegetation, has proven to be very beneficial for certain birds.

The practices that prevent degradation or create environmental improvements on one-year set-aside are more effective, and often less costly, if the same plots are idled for several years in a row. Soil improvements can accumulate and biotopes can develop through their initial stages and support a greater diversity of wildlife and plant species. Negative landscape effects, which are sometimes associated with rotational set-aside, can be reduced. Where trade in set-aside obligations or incentives for taking adjacent parcels out of production creates larger consolidated areas of idled land, the potential for habitat creation and wildlife improvement is further increased. While multiannual set-aside would in principle permit the concentration of idled land in areas where the largest environmental benefits can be obtained, existing programmes have so far provided few incentives for farmers to divert the most environmentally sensitive land. Rather, the least profitable land from an agricultural point of view is placed into short-term set-aside.

There has been a tendency for short-term set-aside to evolve from a pure supply control tool towards a combination of supply control and agri-environmental measure. This development has met with some success, yet it also illustrates the difficulties encountered in "greening" a policy that was originally conceived for supply control, by attaching environmental conditions to it. The integration of environmental concerns into such a policy is constrained by a narrow scope for effective environmental targeting and a limited potential for enduring environmental improvements due to the short diversion period.

In the countries where short-term land diversion has been introduced fairly recently as a policy tool, there is still a lack of experience with managing set-aside. Although land idling had once been part of traditional crop rotations, the agronomic, technical and economic conditions have changed, and today's set-aside has to be managed in ways that are different from the methods of the past. The reintroduction of set-aside into farm production plans provides an opportunity to respond to certain environmental needs related to soil structure, organic matter, and nutrient and water balance, but farmers and policy administrators are having to go through a learning process to ensure that idled land is managed in the environmentally most beneficial way. The approach that will be optimal for a particular farm depends on the local environment and may vary within a country if large differences in natural conditions exists. It is therefore important that the management rules for idled land provide room for regional and local adaptations.

Some land diversion programmes permit farmers to produce alternative crops on set-aside land rather than obliging them to take the land out of production. The environmental impacts of shifting from conventional crops to alternative crops, including those used for the production of non-food materials, are in principle not different from any other shift in production: they can be positive or negative, depending on the crops involved and the production methods applied. The examples discussed in this document suggest that environmental concerns play a subordinate role in the policy decisions affecting alternative cropping. However, the situation differs across countries, and while in the European Union the production of non-food crops is in general subject to less stringent environmental conditions than those applying to other set-aside land, efforts have recently been made in Japan and Switzerland to shift towards soil-improving crops and to encourage the use of conservation practices on set-aside land.

LONG-TERM SET-ASIDE

In general, long-term set-aside can generate considerably larger environmental benefits than short-term land diversion. Since supply control objectives are less important, long-term set-aside can be restricted to lands that are in need of conservation, have a large potential for improvement, or are of strategic importance for habitat creation and natural resource conservation. The contracts signed by farmers can be adapted to local circumstances, and land use practices can be specified that are linked to particular environmental objectives. The financial incentives provided to farmers can be tied to the costs of providing the environmental services or to the value of their benefits.

Among the long-term set-aside schemes, the United States' *Conservation Reserve Program* (CRP) occupies a special position because of its scale, its relatively long history, the number of modifications it has gone through, and the assessments that have been made of its environmental impact. The programme started out as an erosion control measure and was initially targeted at land that had a high erosion potential. Large land areas were enrolled and put under a plant cover during the first years of the programme. With each successive modification, other environmental objectives were added and the criteria for land acceptance were tightened. The new goals reflected a shift away from on-farm environmental concerns towards the impact of agriculture on natural resources that are consumed or valued by the non-farm population. Areas of importance for water quality improvement or for the preservation of certain wildlife habitat, received priority in land enrolment. Recently, there has been a tendency to favour smaller land parcels of high ecological value, such as windbreaks and shelter-belts, filter strips alongside waterways, and agricultural wetlands.

The European Union's long-term set-aside programmes are part of the agri-environmental and forestry schemes introduced with the 1992 CAP reforms. They address a wide range of environmental issues and are currently implemented based on national or subnational programmes. They provide an opportunity for strict environmental targeting and a high degree of regional flexibility. The environmental benefits are likely to be largest where lands of high environmental value are attracted into the programmes and the management practices are adapted to the needs of local ecosystems.

Experience with long-term set-aside indicates that environmental quality can be substantially improved during the contract period. Both the CRP and Canada's *Permanent Cover Program* have reduced soil erosion on arable land and provided benefits for water quality and wildlife. However, there are few guarantees that these benefits will become permanent and persist beyond the duration of the contracts. If the land is eventually returned to production, many of the improvements achieved during the idling period could again be eroded. This concern is increasingly being recognised as more and more long-term set-aside contracts are nearing maturity.

ACHIEVING SUSTAINABILITY

In principle, permanency of environmental improvements will only be ensured if the changes induced by set-aside are environmentally and financially sustainable. There are several ways in which policy can contribute to this goal. One is to favour alternative land uses during the set-aside period that allow farmers to integrate the diverted areas into their farming enterprises, thereby reducing the incentive to revert to previous practices after the end of the contract period. This approach is exemplified by the Canadian *Permanent Cover Program*, which encourages the development of economically viable grass-based uses on former arable land. In this case, paid land diversion aids the transition from one type of land use to another one that is environmentally more desirable. This approach is most promising where the land resources can fairly easily be reallocated to other economically gainful uses.

A second way is to encourage conversion of arable land to woodland, as this provides income in the long run and raises the cost of bringing the land back into cropping after the end of the set-aside contracts. This option will be more attractive to farmers if the length of the set-aside obligation covers the non-productive period of the tree plantings. Experience with the CRP shows that farmers will often choose to reforest areas that were cleared of forests in the not too distant past. Abandoned marginal lands are also candidates for afforestation.

For a large part of the land currently enrolled in long-term set-aside schemes, only limited action has been taken to ensure that the environmental improvements will be sustainable. There remains, of course, the possibility of renewing the contracts indefinitely, but this creates a permanent burden on the budget and will only be justified where there is no other way of securing the environmental benefits. In many cases, less expensive alternatives may exist. By keeping land under a grass cover for an extended period of time, set-aside improves the soil structure and reduces erodibility. To the extent that the land becomes more resistant to environmental pressure, it can be brought back into production, provided that certain cropping practices are used. The *Conservation Compliance Provision* in the United States, which obliges farmers to use conserving practices on erodible land, would protect much of the CRP land if this land were returned to production. In the European Union, incentives have recently been introduced to prolong the effective duration of set-aside by placing land coming out of short-term set-aside into longer term schemes. Farmers will generally find it easier to engage in long-term land use planning if arrangements aimed at maintaining the environmental benefits of set-aside beyond the contract period are specified before the contracts are signed.

Set-aside has reduced the land base available for crop production and, in areas with scarce land resources, has increased the competition for land. Whether this has affected application rates of fertilisers and plant protection products on the land remaining in production and increased the environmental pressure on this land, is yet unclear, although preliminary evidence suggests that there has been little change in the intensity of fertiliser and chemical use in response to land diversion. In cases where the price relationships determining input use have changed, it would be expected that set-aside may have led to a substitution of fertilisers and farm chemicals for land in production. On the other hand, farmers with a fixed labour base and no off-farm employment opportunities may have resorted to more labour-intensive tillage practices, with less intermediate input use, including fertiliser and farm chemicals.

SUPPLY EFFECTS

With respect to reducing the supply of arable crops, rotational set-aside is likely to be the most effective form of land diversion. All other forms allow farmers to retire the least productive land and can be associated with supply reductions that are considerably smaller than the proportion of land idled. However, one-year set-aside can generate substantial improvements in soil fertility and yield increases in subsequent years, in which case the supply-reducing effect of rotational set-aside may not be much larger than that of other forms of land diversion. Of the pro-grammes discussed in this document, only the European Union's arable payment scheme included a rotational set-aside obligation when it was first introduced. Shortly after its implementation, however, farmers were offered a non-rotational option to satisfy their set-aside requirement, which is increasingly being used.

Regarding long-term environmental set-aside, there has been a tendency since 1995 to permit participating farmers to terminate their contracts before the date of expiration. This development, which has been motivated by the high world market prices for grains, will free some of the set-aside land for production. However, it also illustrates the trade-off between supply control and environmental objectives, and raises the question as to how far the environmental restrictions should be relaxed to allow farmers to take better advantage of favourable market conditions. A large-scale return of land enrolled in long-term set-aside schemes to production without appropriate environmental safeguards could obliterate many of the environmental benefits that were achieved at taxpayers' expense, and undermine the credibility of environmental set-aside.

BUDGETARY COSTS

The set-aside programmes discussed in this document all provide a financial incentive to farmers in the form of direct payments for idled land, or by making set-aside a condition for receiving payments under related farm programmes. None of the land diversion schemes obliges farmers to bear the full economic cost of setting aside land. The per-hectare payments vary substantially across programmes and countries. In the case of environmental set-aside, they often depend on site-specific characteristics and on the type of environmental management restriction applying to the land.

Under the Canadian *Permanent Cover Program*, a one-time payment of between C$ 50 (US$36)[2] per hectare and C$ 163 (US$119) per hectare was made, depending on the length of the contract. In addition, the government made a financial contribution towards the cost of seeding of C$ 50 (US$36) per hectare. The annual payment rate for set-aside under the European Union's arable aid scheme is specified in ECU per tonne and has to be multiplied by the reference yield for cereals applicable to the region to obtain the per-hectare payment. The payment rate for the 1995/96 marketing year was ECU 69 (US$90) per tonne. Annual per-hectare payments for long-term set-aside under the agri-environmental regulation vary from ECU 146 (US$191) to ECU 600 (US$784). Under the forestry scheme, payments of up to ECU 725 (U$947) per hectare and year are allowed during the non-productive period of tree plantings.

In the United States, the average annual rental payment for land signed up in the *Conservation Reserve Program* between 1986 and 1992 was US$125 per hectare. In Japan, the average subsidy for rice paddy field diversion was ¥ 108 000 (US$1 056) per hectare in 1994. Under the current programme, Japanese farmers receive a base payment of ¥ 70 000 (US$740) for the most widespread form of set-aside (planting of annual crops on diverted paddy fields). This payment is increased if farmers take measures that encourage a more efficient use of natural resources. In Switzerland, payment rates range from SF 300 (US$254) per year and hectare for low-intensive pasture land to SF 3 000 (US$2 537) per hectare for green fallow and some other measures.

Since participation in the land diversion programmes is voluntary, the payments have to be at least as high as the income foregone by taking the land out of production and complying with the environmental conditions. A high level of price support raises the income foregone and is usually reflected in high per-hectare payments for diverted land. Total budgetary expenditures on land diversion schemes still account for only a few per cent of agricultural support as measured by the PSE, but in some countries they are among the fastest growing types of direct payments to farmers.

The payment level is an important determinant of the environmental effectiveness and the cost-efficiency of the programmes. For environmental set-aside, the challenge is to establish the "price" of an environmental service for which there is a public demand but no market, and the production of which requires farmers to forego income. A competitive bidding process, by which farmers specify the minimum payment level they are willing to accept in return for setting their land aside, could ensure that the costs are not overstated. Ranking of the bids based on expert assessment of the expected environmental benefits would guarantee that offers with the largest benefits are considered first for acceptance. A combination of the two may go some way towards an efficient allocation of land resources and budgetary funds. The CRP is the only programme so far that has adopted this approach.

However, the payment system that is most suitable for a particular situation has to strike a balance between the gains from and the extra administrative effort involved in achieving a more efficient resource allocation, including the costs of building up the expertise and information required to differentiate payment rates. Although uniform rates are not likely to lead to the most efficient resource allocation, they may be the only practical alternative at present in countries that have never operated similar schemes before. In such cases, efforts should be made to develop the administrative infrastructure necessary to improve the targeting of the programmes. In this context, use of new technologies can contribute to reducing the administrative burden of land diversion schemes. In the European Union, for instance, remote sensing techniques have been developed to facilitate programme implementation and monitoring.

A competitive payment system also faces the difficulty that the environmental services obtained from diverting land are site-specific, which could entice farmers to include a "monopoly mark-up" in the bid price. Nevertheless, estimates in the United States suggest that by moving towards a competitive bidding process and a ranking of the bids based on their environmental value, the benefit/cost ratio of the CRP has been increased.

THE CHANGING ROLE OF SET-ASIDE

Land diversion can only address certain types of environmental issues. Some ecological systems and landscapes of high value, which are the result of traditional low-input agriculture, would disappear if the land were idled and the agricultural activity discontinued. Other goals, such as watershed protection, may be better addressed by "zonal" programmes; and to respond to changes in consumer preferences based on the way food is produced, a whole-farm approach may be required. To the extent that land diversion can help address these concerns, it will be able to do so only in combination with other measures, especially those that encourage the adoption of less intensive cropping practices and integrated farming systems. In some countries there has been a development from land diversion as a supply control tool or a stand-alone measure for environmental conservation, to mixed programmes in which set-aside is one of several policy components. This development has progressed particularly far in Switzerland, but it is also noticeable in several EU countries.

Environmental monitoring of the land diversion schemes is generally inadequate, partly because of the recent origin of the programmes. Knowledge of the economic and natural factors that determine the environmental impacts of the schemes is still limited. At best, model-based estimates are used to predict the effects on one or two key environmental parameters, such as soil erosion. *Ex post* evaluations of the true environmental impacts are only about to begin. Current efforts to develop agri-environmental indicators, especially those measuring regional and site-specific effects related to agricultural land and input use, and the state of natural resources, including land, water, ecosytems and wildlife habitat, but also landscape, could be useful in this respect. They could help improve the design of future programmes and provide the feedback necessary for increasing the effectiveness of current schemes. Regular evaluation of the environmental impacts could provide a basis for verifying that the taxpayer funds expended on land diversion schemes are directed to their most effective use.

The number of environmental measures in agriculture is rapidly increasing. In the future, a greater effort will have to be made to ensure that land diversion schemes are well integrated and consistent with these measures. Moreover, the objectives pursued by land diversion policies might change, as some of their current functions may be performed by other programmes. Erosion control could thus be taken over, at least on part of the erodible land, by initiatives that encourage conservation practices in cropping.

Land diversion programmes have often been implemented in the context of a high level of market price support, which is likely to have increased use of fertilisers and pesticides, drawn marginal land into production, and created pressure on the environment. The current land diversion programmes may to some extent merely undo some of the production and environmental effects of past and current price support.

In countries in which support to agriculture is still largely linked to commodity production, agricultural policy reforms, centred on reductions in institutional prices and a shift from price support to direct payments, could alter the role of land diversion schemes. A lowering of price support could reduce the need for supply control and thus for short-term set-aside schemes. Policy reforms could also lower the pressure on the environment and, in combination with the introduction of targeted agri-environmental measures, free long-term set-aside of some of its current objectives. Long-term land diversion schemes could then be restricted to situations where they remain the most effective and cost-efficient means of providing environmental benefits demanded by society.

THE ENVIRONMENTAL EFFECTS
OF AGRICULTURAL LAND DIVERSION SCHEMES

1. OBJECTIVES OF THE STUDY

The main purpose of this study is to improve the understanding of the *environmental impacts* of agricultural land diversion schemes in OECD countries. This work is part of the larger effort of the JWP to: *i)* explore the linkages between agricultural policies and the environment, and *ii)* identify the environmental effects of agricultural policy reforms.

For the purpose of this analysis, the term *land diversion scheme* refers to policy measures that require or pay farmers to take land out of agricultural production, or result in land use changes. By restricting the area that would otherwise be farmed, such schemes are directly linked to the management of farmland, with consequences for agricultural production and input use, and the quality of the environment.

Land diversion programmes form part of the wider range of measures limiting the supply of agricultural commodities in OECD countries. Yet, they are also used to integrate environmental concerns into agricultural policy. In practice, land diversion schemes are often implemented with the aim of pursuing both supply control and environmental objectives, and have been modified over time to reflect changes in the relative importance of each of these objectives.

Although the primary goal of the analysis is an assessment of the environmental impacts of land diversion schemes, the budgetary cost of these schemes and their interactions with other policy measures in agriculture are also addressed. Since land diversion programmes are often used to curb surplus production that would not arise in the absence of agricultural support, and to address environmental pressures that are sometimes increased by support, the linkage between land diversion schemes and agricultural support policies is of particular relevance.

A further objective of the analysis is to identify the extent to which currently existing data, including agri-environmental indicators, can be used to contribute to the evaluation of the environmental impacts of land diversion schemes. One element in this assessment is the degree to which agri-environmental relationships are specific to regional or local conditions. The insights gained in this area could contribute to advancing the work on agri-environmental indicators in the OECD and in Member countries.

The study is organised as follows: in Section 2, the framework of analysis is described. In Sections 3 through 7, the land diversion schemes are analysed, with each of the sections corresponding to one of the five countries whose programmes are being examined.

2. FRAMEWORK OF ANALYSIS

Environmental changes result from several factors. These include price and income support policies in agriculture, which can affect farm management practices, the location of production, the mix of commodities produced and the intensity of agricultural resource use; shifts in relative prices; technological and biotechnological innovations; and changes in attitudes and in the environmental awareness of producers and consumers. The impact of land diversion schemes can be seen as causing additional changes in the environment, interacting with those that would have taken place due to the other factors.

Given these multiple and concurrent factors, it is necessary to define the most appropriate alternative situation ("baseline") against which the environmental effects of land diversion schemes should be evaluated. In accordance with the approach used in related OECD work, and given the exploratory and pragmatic nature of the analysis, the approach followed in this study is to compare the environmental situation before and after implementation of the land diversion schemes. Where possible, estimates of environmental changes that can be attributed to economic, technological or natural factors or changes in other policies, are accounted for in the analysis in order to better separate the effects of land diversion programmes from those of other factors.

While there may be shortcomings to this approach in cases where numerous factors have contributed to the environmental changes, this may be less of a problem in this particular study because many of the environmental benefits of land diversion programmes are obtained on set-aside land, where they can be more easily attributed to the policy measure than on land that remains in production: except for the impact of natural events and deposition or infiltration of polluting substances from the surrounding environment, changes in the environmental quality of set-aside land are mainly due to the land diversion programme itself. By comparison, environmental impacts on the land that remains in cultivation could also be caused by other factors.

The land diversion programmes have been examined at three levels. First, the criteria with respect to farmer participation and land eligibility, and the financial incentives provided to farmers are discussed. While this does not address the actual environmental impacts of a programme, it can provide some insights into the degree of environmental targeting and can be helpful in cases where no empirical information on the environmental effects of the programme is available. Second, the changes in land use, input use and farming practices associated with land diversion schemes are assessed. This includes an analysis of the physical character- istics of the diverted land and the changes in cropping patterns and management practices. Finally, data on the physical, chemical and ecological effects of the schemes with respect to the quality of soils, water, air, biodiversity, wildlife habitat and landscape are analysed. The data used in the analysis are collected from various sources, and include official statistics, environmental indicators, results of scientific studies and model-based estimates.

In addition to the environmental impacts of the land diversion schemes, the paper examines the effects of the programmes on commodity supply; the budgetary implications of land diversion payments; and the relationship between land diver- sion programmes and other agricultural and environmental policies.

3. CANADA'S PERMANENT COVER PROGRAM

Background

Canadian land diversion policy dates back to the mid-1930s, when the *Prairie Farm Rehabilitation Act* was introduced in response to severe drought in the west- ern provinces. The Prairie Farm Rehabilitation Act was aimed at removing land from grain production and putting it to pasture uses. The *Permanent Cover Program* (PCP) was introduced in 1988 with the objective of reducing soil deterioration on high-risk land and improving wildlife habitat. Under the PCP, farmers are encouraged to take farmland out of annual cultivation and convert it to perennial cover for periods of either 10 or 21 years. The programme is targeted at highly erodible and other environmentally sensitive land where annual cropping is causing soil degradation.

Implementation of the PCP began in 1989 and was initially limited to the provinces of Alberta and Saskatchewan ("PCP I"). Around 360 000 acres (1 acre = 0.4 hectares) of marginal and erodible land were enrolled under PCP I. In 1991, the programme was modified and extended to include Manitoba and the Peace River Region of British Columbia ("PCP II") and, at the same time, was closely integrated with the *North American Waterfowl Management Plan*.[3] Land was accepted into the PCP until 1993 when new enrolment was halted. Overall, about 1.3 million acres of cropland were diverted from grain production under PCP I and PCP II.

The programme

The PCP authorised the government to grant a one-time payment to farmers for taking highly erodible or otherwise degradable land out of grain production and converting it to long-term cover such as forage, pasture or forests. It also authorised the government to contribute to the cost of establishing the cover. Land eligibility for the PCP was largely determined based on land inventory maps, but other information, including soil survey maps, aerial photographs and crop insurance data, were also used. Land considered marginal by soil scientists (Canada Land Inventory classes 5 and 6) was automatically eligible under PCP I; class 4 land became eligible under PCP II. Other environmentally sensitive land needed additional assessment before it could be enrolled in the programme. A land parcel qualified for the PCP if at least half of it was eligible land and if this area was at least 40 acres in size (Gray and Paddock, 1993).

About 81 million acres of farmland were used for annual cropping, including summer fallow, in the area covered by the PCP at the time when the programme was introduced. More than 12 million acres of this land were classified as being highly erodible (Canada Land Inventory classes 4, 5 and 6). This land, most of which is located in the provinces of Alberta and Saskatchewan, constituted the primary target area for PCP enrolment. In addition, two million acres of cropland affected by high levels of soil salinity and almost 12 million acres of land with acidic soils or wetland conditions, were deemed suitable for protection (Prairie Farm Rehabilitation Administration [PFRA], 1992).

Farmer participation in the PCP was voluntary. Participants signed a contract for 10 or 21 years, committing themselves to seeding the targeted area and maintaining the permanent cover for the specified time. Agricultural uses of the land, such as grazing, are permitted as long as they are compatible with the permanent cover requirement. In return for accepting the restrictions on land use, farmers received a financial contribution towards the cost of seeding of C$ 20 (US$15) per acre and, once the viability of the permanent cover was established, a one-time payment. At the beginning, a bidding process was used to determine the size of the payment, but later it was replaced by fixed amounts of C$ 20 (US$15) or C$ 30 (US$22) per acre for 10-year contracts, and C$ 50 (US$36) or C$ 65 (US$47) per acre for 21-year contracts (PFRA, 1992). Total payment per landowner was limited to C$ 64 000 (US$46 600) (Gray and Paddock, 1993).

The PCP is administered by the regional and district offices of the Prairie Farm Rehabilitation Administration. Soil conservation specialists provide technical assistance and agency administrators are charged with ensuring that the programme conditions are met by farmers.

Around 15 000 farmers enrolled in the PCP, removing some 1.3 million acres of farmland from crop production. About four-fifths of total PCP land was signed up in the provinces of Alberta and Saskatchewan. The programme was fully subscribed in the sign-up periods, and there have been repeated requests from both farmers and environmental groups for a continuation of the PCP. The broad acceptance of the PCP is credited to the fact that the programme was developed in close consultation with producers and environmental groups, and addresses the concerns of the different stakeholders.

Environmental assessment

The PCP is directed at land where annual cultivation is causing long-term soil damage and where special farming practices are not sufficient to eliminate the risk of deterioration. By putting this land under a perennial or permanent cover, it is intended to conserve or improve soil productivity and, at the same time, generate benefits for water, air, wildlife habitat and landscape.

The PCP has to be assessed against the background of overall trends in land use, farming practices and soil quality. Although the area of cropland has increased by about 2.5 million acres between 1981 and 1991 and much of the marginal land that can be used for agricultural production is under cultivation (Agriculture Canada, 1995), there is evidence that changes in land use and farming practices have in some areas led to improvements in soil quality over the last decade. Adoption of conservation tillage, improved residue management, extended crop rotations, increases in forage cover and a reduction in summer fallow are the changes that are likely to have had the greatest benefits to soils. Conservation tillage is now used on roughly two-thirds of the prairie farmland.

Assessments show that these changes have lowered the risk of wind and water erosion, salinisation and depletion of organic matter. It has been estimated, for instance, that the risk of wind erosion in the Prairie Provinces has been reduced by seven per cent in the period from 1981 to 1991 (McRae *et al.*, 1995). Roughly two-thirds of these improvements have been attributed to changes in tillage practices, the rest to changes in cropping systems. Similarly, the risk of water erosion in the prairies is estimated to have declined by between 8 and 15 per cent. Although at this aggregate level it is impossible to determine how much of these improvements are due to the PCP, it is likely that the shift of prairie land from grain production to forage-based uses under the PCP has helped conserve soil. However, several factors are at play and the increasing use of conservation practices on cropland may have reduced the environmental benefits of land diversion.

Assessment of the land diversion criteria

Land eligibility for the PCP was determined based on Canada's Land Inventory classification system, supplemented by other scientific information. There is widespread agreement in Canada that the classification systems established by soil scientists provide a good basis for identifying the land that is unsuitable for annual cropping and should be kept under a permanent cover. However, because of the limited funds authorised for the PCP, only a relatively small proportion of the eligible cropland was effectively enrolled in the programme. In fact, there is evidence that the rapid uptake during the sign-up periods reduced the opportunities for late adopters to participate in the PCP (PFRA, 1992). This raises the question of whether the land taken out of crop production was also the most environmentally sensitive land and, given the uniform payment rates, whether the changes have been achieved at the lowest budgetary cost. To answer these questions would require a more detailed analysis of the land selection process and the environmental characteristics of the land enrolled in the programme.

Effects on land use, input use and farming practices

According to a 1994 survey, more than half of the farmers participating in the PCP have mixed farming operations, and roughly one-third are livestock farmers (Western Opinion Research, 1994). Only four per cent of the respondents operate a grain farm. The majority of the lands enrolled in the PCP are used for grazing, hay production and, to a lesser extent, forage seed production. The forage cover was established over the period from 1989 to 1993, with the major efforts taking place in 1991 and 1992.

Haying is the most common activity on PCP land. Four-fifths of the surveyed participants use at least part of their PCP land for haying. In most cases, hay is cut once a year. Almost two out of three farmers have livestock grazing on their PCP. Grazing is most widespread in the brown soil zone and on land that is under 21-year contracts. Most of the grazing activity takes place in the autumn season and involves beef cattle, yet horses, sheep and dairy cattle can also be found. The average size of PCP forage land per farm is 137 acres; land parcels used for grazing extend over almost 200 acres on average (Western Opinion Research, 1994).

With regard to farming practices and input use, only about one-third of the respondents applied commercial fertilisers, 20 per cent spread manure and 10 per cent used pesticides on PCP land. Considerable differences in the use of farm chemicals were found across regions. In Manitoba, for instance, commercial fertilisers were applied in almost 75 per cent of the cases. Overall, however, use of farm chemicals is substantially lower on PCP land as compared with land under annual cultivation.

Because of the limited duration of the PCP contracts, there is a possibility that the land could be returned to grain production at a later date. In some cases, the soil improvements achieved on the diverted land might permit subsequent cropping on a sustainable basis, provided that conservation tillage and other conserving practices are applied. However, much of the land might have to remain in forage or trees to ensure that the environmental improvements are not lost after expiration of the PCP contracts (Tyrchniewicz and Wilson, 1994). When asked about future plans for their PCP land, a large majority of farmers indicated that they would keep diverted land in forage as long as possible, and more than half had plans to seed additional cropland to forage. Fewer than 20 per cent of the respondents had intentions to return PCP land to grain production. Farms with 21-year contracts, livestock enterprises and farms using PCP land predominantly for grazing are least likely to shift diverted land back to annual cultivation (Western Opinion Research, 1994). Furthermore, no attempts have been made by farmers to opt out of their contracts and return the land to grain production in response to the recent increases in grain prices.

Effects on the environment

Since quantitative indicators for monitoring the environmental impacts of the PCP have not been developed, assessment of the environmental effects of the programme depends upon other sources of information, including farm survey results, occasional scientific studies, and general information about the impact of forage and pasture uses of prairie land on environmental quality.

The principal objective of the PCP is to control *soil erosion*. Both the land eligibility criteria and the permanent cover requirement are consistent with this objective. The two predominant activities on PCP land, forage production and extensive grazing, are among the most conserving land uses on erodible land. Moreover, the majority of the PCP contracts were signed in Alberta and Saskatchewan, the two provinces with the largest areas of highly erodible land. It is therefore not surprising that the effect of the PCP most commonly mentioned by programme participants is a decrease in the amount of soil erosion. Overall, 74 per cent of surveyed participants reported a decline in soil erosion. By comparison, only one-third of the respondents perceived improvements in *water quality* (Western Opinion Research, 1994).

Another beneficial effect of the PCP is likely to result from an increase in *organic matter* on diverted land. Long-term studies have shown that the conversion of rangeland to cropland is associated with a substantial diminution of organic matter (Agriculture Canada, 1995). Conversely, establishment of forage cover on erodible land is likely to lead to a gradual accumulation of organic matter, a greater diversity of biological soil organisms and an improvement in soil texture. These changes are important elements for restoring *soil productivity*.

In addition, the permanent cover slows down absorption of water into the soil and reduces evaporation, thereby driving surface salts deeper into the ground and lowering the level of salinity in the upper soil strata (Saskatchewan Wetland Conservation Corporation). The 1994 survey results indicate that one out of three farmers perceived reductions in *soil salinity* and that these were positively related to the successful establishment of forage cover. Improvements with respect to soil salinity appear to be more widespread on grain farms than on livestock operations, and most noticeable on brown and dark brown soils (Western Opinion Research, 1994).

The Canadian prairies are an important breeding ground for North American birds. Since the 1950s, the prairie bird population has sharply declined (by two-thirds in the case of breeding ducks), and several wildlife species have become extinct or endangered. The increase of grass and forage cover due to the PCP has created favourable conditions for wildlife development. The population densities of most grassland songbirds were found to be higher on permanent pasture than on cropland. Large tracts of pasture land also make for an excellent duck nesting habitat, allowing ducks to space nests widely and to hide from predators. The ample size of the plots enrolled in the PCP provide effective wildlife travel lanes and increase wildlife ranging areas. In 1994, two-thirds of surveyed PCP participants reported an increase in wildlife habitat as a result of the programme.

Although the effects of the PCP on the environment have largely been positive, some negative impacts have also been identified. Almost half of the surveyed farmers reported problems with rodents on their PCP land, and some experienced difficulties with perennial weeds and grasshoppers (Western Opinion Research, 1994). Competition from wildlife was mentioned by 16 per cent of the respondents, although the competition between domestic and wildlife species for forage could also be interpreted as a success for wildlife habitat creation.

Budgetary expenditures, economic effects and relationship with other farm programmes

Total federal government payments to producers under the PCP amounted to around C$ 70 (US$51) million. The average cost per acre was C$ 57 (US$42) in the case of 21-year contracts. The administrative costs of the programme were estimated to correspond to roughly one-quarter of the payments made to farmers (Gray *et al.*, 1993).

A move towards environmental sustainability is dependent in the long run on agricultural activities also being financially sustainable. In this respect, the PCP could be seen to be successful on both accounts. In the 1994 survey of participants, 70 per cent reported a decrease in operating costs, and 56 per cent reported an increase in net farm income. Only seven per cent saw their net income decline. Increases in net farm income were especially widespread among livestock farmers and farms that had established the forage cover early after introduction of the programme. The latter suggests that the economic benefits of the land use changes materialise to the full extent once the farm management adjustment is completed. While PCP payments enter into the income calculations, there is some indications that even in the absence of subsidies there would still be a net benefit from the changes in land use (Tyrchniewicz and Wilson, 1994). If this is the case, the financial incentives provided by the PCP would have facilitated a shift from grain production to forage-based land uses that are economically as well as environmentally sustainable.

The relationship between the PCP and other farm programmes is examined here from two angles: *i)* the implications of the PCP for the budgetary costs of other farm programmes; and *ii)* the effects of other farm programmes on the environment.

With respect to government expenditures on agricultural support programmes, the PCP is thought to have resulted in net savings for the federal budget because of its effect on grain production and associated subsidy reductions in the grain sector. According to an estimate based on 1992 data, every dollar spent on 10-year PCP contracts resulted in budgetary savings of 2.4 dollars. For 21-year contracts, budgetary savings were estimated at three times the programme expenditures (Gray et al., 1993). In 1993, the total budgetary savings due to the PCP were estimated at C$ 28.9 (US$21) million per annum.[4] However, these savings were calculated based on the government programmes in place at the time, and government subsidies for grain production have since then declined.

With respect to the effects of agricultural support on the environment, there is reason to believe that farm programmes in general have rewarded expansion of cropping onto marginal soils and rangeland that is unsuitable for annual cultivation (Girth, 1990). By supporting diversification from cropping into forage production and grazing, the PCP may to some extent have offset the environmental effects of previous agricultural support policies.

Analysis of individual farm support programmes suggests that their effects on the environment can be quite complex. An evaluation of the Western Grain Transportation Act, a programme that until 1995 authorised government payments to the railways to compensate for the movement of grains from the prairies to shipping ports, came to the conclusion that the programme may have reduced summer fallow and livestock intensity in the western provinces, with positive consequences for soil and water quality, but negative effects for wildlife because of a decrease in forage area and increased input use on cropland[5] (Terrestrial and Aquatic Environmental Managers, 1992).

Two other important Canadian farm programmes are the Net Income Stabilisation Account (NISA) and the General Revenue Insurance Program (GRIP). Under NISA, farmers are encouraged to stabilise income by paying into individual NISA accounts in good years and withdrawing from these accounts in bad years. Producers can deposit up to 20 per cent of annual net sales in the account. The first two per cent are matched by a financial contribution from the federal and provincial governments. GRIP provides payments to producers when revenues from crop production fall below a specified target level. The insurance premiums are co-financed by producers, provincial governments, and the federal government (Agriculture Canada, 1993).[6]

Assessments of the GRIP and NISA programmes suggest that these schemes might have contributed to soil degradation due to a conversion of marginal lands to crop production (Agriculture Canada, 1993). The risk-reducing effect of the programmes may also have led to a decline in cropping and land use diversity. On the other hand, the schemes have reduced the summer fallow acreage and increased the use of conservation tillage, with potential benefits for the environment. No impact on the intensity of fertilisers and pesticide use was found. Overall, the environmental effects appear to have been rather small in magnitude. The most significant impact appears to have been a conversion of wetlands to cropland (Environmental Management Associates, 1993).

Concluding remarks

The discussion of the PCP suggests that the programme has led to the conversion of a substantial area of environmentally sensitive prairie land from grain production to forage and livestock-based uses. Among the beneficial environmental benefits of these changes are a reduction in soil erosion, an increase in soil productivity, improvements in the quality of surface water, and the preservation and creation of bird, fish and other wildlife habitat. The positive environmental effects of the PCP are linked to the targeting of erodible land, the substantial size of the diverted plots and the permanent cover requirement. Moreover, the benefits seem to have been achieved at a reasonable budgetary cost.

However, given the large area of prairie cropland classified as highly erodible and degraded by annual cultivation, the achievements of the PCP would seem to be rather limited in scope. Only slightly more than 10 per cent of the eligible land was enrolled in the programme. The PCP may also have led to possible increases in erosion on the land remaining in crop production due to more intensive cultivation on that land, but this has yet to be established. Furthermore, there is some indication that environmentally vulnerable land is not always marginal in economic terms and consequently, farmers may have been reluctant to divert environmentally sensitive yet profitable land from crop production.

In this light, it is possible that the environmental effectiveness and cost-efficiency of the PCP could have been improved through a more selective land acceptance process and differential compensation payments. One way of achieving this might have been by inviting farmers to bid for contracts rather than granting a fixed per-acre payment. Also, farmers' bids could have been evaluated based on the expected environmental benefits from land enrolment, taking account of the environmental objectives of the programme and the seriousness of the environmental problems in a given area. Such modifications might have made programme participation more attractive in areas with high expected environmental benefits and a low opportunity cost of land use.

Although there is no guarantee that the environmental achievements of the PCP will be maintained beyond the duration of the contracts, it appears that under current market and policy conditions, most of the PCP land will not be returned to annual cultivation. Most participants have mixed farming or livestock operations and are therefore able to integrate forage production or grazing on the PCP land into their long-term farm management plans. Thus, the land use changes initiated by the PCP will in many cases become economically sustainable. On the other hand, the low participation of grain farmers in the PCP raises the question of whether the programme's bias in favour of farms engaged in livestock production has not effectively excluded environmentally sensitive land on grain farms, which can be particularly vulnerable to wind erosion, from being enrolled.

4. THE EUROPEAN UNION'S LAND DIVERSION PROGRAMMES

The land diversion schemes addressed in this section include the five-year set-aside scheme introduced in 1988 (Council Regulation 1094/88), the set-aside requirements for farmers who wish to claim compensatory payments under the arable aid scheme (CR 1765/92), long-term land diversions for agri-environmental purposes (CR 2078/92), and the forestry aid scheme (CR 2080/92).

Background

In 1988, the EU introduced a *voluntary five-year set-aside* scheme that paid producers to divert cropland from production for a period of five years (CR 1094/88). The programme was intended to reduce production and contribute to environmental conservation. Participating farmers had to set aside at least 20 per cent of their arable land. In return, the received a payment that compensated for the income foregone by taking the land out of production.

In 1991, a *temporary set-aside* scheme was introduced in parallel to the voluntary five-year scheme (CR 1703/91). Land withdrawn under this scheme had to be kept out of production for one year, put under a plant cover, and managed according to certain environmental requirements.

With the reforms to the Common Agricultural Policy (CAP) in 1992, the European Union made set-aside a condition for receiving government support on arable land. Farmers producing more than 92 tonnes of arable crops ("commercial farms") were obliged to idle a specified percentage of their arable land each year (*market set-aside*) in order to receive area-based support payments for cereals, oilseeds, protein plants and linseed. The old five-year set-aside scheme was closed to new entrants.

In addition, two *long-term land diversion schemes* were introduced in 1992 as part of the "accompanying measures" of the CAP reforms. These two schemes are specifically aimed at achieving environmental objectives. The first is part of the *agri-environmental regulation*, a policy package for the promotion of agricultural production methods compatible with the protection of the environment and the maintenance of the countryside (CR 2078/92). Under this regulation, farmers can receive payments for diverting land from production for environmental purposes for a period of at least 20 years.

The second long-term land diversion scheme is aimed at supporting the development of farm forestry as an alternative to agricultural land use (CR 2080/92). This programme provides payments for the afforestation of agricultural land and for the development of forestry activities on farms. The 1992 *forestry scheme* replaced previous measures which, according to the European Commission, did not provide sufficient incentives for afforestation.

The programmes

The 1988 voluntary **five-year set-aside** programme was a multi-purpose scheme. It was intended to reduce cereals plantings and decrease the budgetary expenditures associated with storage and disposal of surplus grains. At the same time, it was used to encourage farmers to diversify production, shift to extensive grazing, create wildlife cover, and provide other environmental benefits. The size of the payment was determined by Member states, but could not exceed ECU 732 (US$957) per hectare and year (Communautés européennes, 1995). The scheme was financed jointly by the Union and Member country governments. Basic requirements with respect to the management of fallowed land were specified at the Union level, but Member countries were given considerable discretion in defining additional environmental conditions. Management rules included a ban on pesticides and herbicides, the need to establish a plant cover and rules for extensive grazing.

Market-set aside was introduced primarily for the purpose of supply control. To be eligible for support payments under the EU's arable aid scheme, commercial farmers are obliged to set aside a certain percentage of a base area. The base area is equal to average plantings of cereals, oilseeds and protein crops over the three-year period from 1989 to 1991, including land fallowed under a government scheme during that time (Commission of the European Communities [European Commission], 1993a).[7] In principle, base areas can be defined either for individual farms or for regions, but in practice all Member states opted for regional base areas (European Commission, 1995a). The set-aside percentage is determined annually by the Agriculture Council based on the market situation for grains (taking account of developments in supply, demand, exports and stocks), the objectives of the reformed CAP, and budgetary considerations.[8] Set-aside parcels must have a minimum size of 0.3 hectares and a width of at least 20 metres.[9]

Market set-asides can be managed in a rotational or a non-rotational way. Under the *rotational* option, a certain percentage of the land for which farmers claim arable area payments has to be idled in a given year, and the same plot of land can not be set aside more than once in six years. *Non-rotational* set-aside permits farmers to divert a given parcel of land for as long as they choose. A special variant of non-rotational set-aside is five-year or *guaranteed set-aside*, which offers the additional benefit to farmers that the per-hectare payment can not be decreased during the contract period. In addition to their set-aside obligation, producers may idle, at the same payment rate, arable land on a voluntary basis (*voluntary set-aside*). The total diverted area including voluntary set-aside can not exceed 50 per cent of the area for which arable area payments are claimed; individual Member states may set lower ceilings.[10]

The initial set-aside requirement for the *rotational* form was 15 per cent. The *non-rotational* option, which became available in the 1994/95 marketing year, originally required farmers to set aside at least 20 per cent of the land (18 per cent in the United Kingdom and Denmark). For 1995/96, the set-aside requirements were reduced to 12 per cent for rotational and 17 per cent (15 per cent in the UK and Denmark) for non-rotational set-aside; for the 1996/97 marketing year, the percentages were lowered to 10 per cent for both types of set-aside, and for 1997/98 they have been further reduced to 5 per cent, reflecting a tighter market situation for cereals.

If total plantings of cereals, oilseeds, linseed and protein crops in a region exceed the regional base area, a penalty is imposed by reducing payment levels for the current year, and increasing set-aside percentages for the following year (*penalty set-aside*). No payments are made for penalty set-aside. Since all producers in a reference region are equally affected by the penalty irrespective of their contribution to the overshoot, there can be an incentive for individual farmers to maximise the area under the scheme by bringing additional land into crop production. To preclude this possibility, acreage payments can generally not be claimed on land that was planted to permanent grass, permanent crops or trees, or which was used for non-agricultural purposes at the end of December 1991.

The payment rate for set-aside land is specified in ECU per tonne, and has to be multiplied by the reference yield for cereals applicable to the region to obtain the per-hectare payment for land in this area.[11] The payment rate for 1995/96 was ECU 69 (US$90) per tonne; by comparison, the 1995/96 intervention price for cereals was ECU 119 (US$156) per tonne (Communautés européennes, 1995). The payments are entirely financed by the European Community budget.

As an alternative to setting aside land, commercial farmers can choose to claim arable payments only for an area of land that would produce less than 92 tonnes of cereals equivalent ("simplified scheme"). In this case, no set-aside is required and the payments for all eligible crops on this limited land area are based on the payment rate for cereals.[12]

The diverted land has to be maintained in a manner that ensures protection of the environment. Within general guidelines established at the Union level (CR 2293/92 and CR 762/94), Member states can specify detailed management rules concerning plant cover, use of fertilisers and farm chemicals, and spreading of animal manure. They can, for instance, draw up a list of permitted cover crops, restrict cuttings of the grass to certain time periods, and specify measures for weed control (European Commission, 1993*b*).

No agricultural use is to be made of set-aside land, except for the production of raw materials for industrial use ("non-food crops"). Such crops may be grown without a reduction in the set-aside premium, provided that the value of the industrial product is greater than the value of the food and feed by-products obtained during processing. In the case of annual crops, a contract has to be signed between the farmer and a processor before planting time. Permitted non-food raw materials include cereals, certain oilseeds and protein crops, trees with a harvest cycle not exceeding 10 years, and certain shrubs and bushes. Eligible non-food uses include biofuels, biomass, industrial oils and fats, plastics, paper, chemicals and pharmaceuticals. Sugar beet production for industrial purposes is also allowed, but in this case no area payment is made. Land used for the production of non-food crops is exempt from the environmental management restrictions normally applicable to diverted land.

Since the introduction of the arable aid scheme, a number of modifications have been made to the rules governing market set-aside. Since the 1994/95 marketing year, trade in set-aside obligations within 20 kilometres of the farm or for the purpose of transferring set-aside to regions in which specific environmental objectives are pursued, can be allowed by Member states.[13] In the case of a transfer, the set-aside obligation is increased by 5 percentage points (3 percentage points in the case of the UK and Denmark). Concerning excess plantings, a provision was introduced in the 1996/97 marketing year that allows 85 per cent of the land entered into *voluntary set-aside* to be deducted from the area exceeding the regional base area before the *penalty set-aside* requirement is calculated, thus reducing the cost of non-compliance with the regional planting limits.

Until 1995/96, termination of *guaranteed set-aside* before the end of the 5-year period was subject to a penalty, although a shift of land from guaranteed set-aside to the agri-environmental or forestry scheme was possible. For 1996/97, farmers were allowed to terminate their five-year contracts without a penalty. At the same time, new contracts for guaranteed set-aside were limited to farmers who join an access scheme or a national environmental scheme under CR 2078/92, or who plant a fast-growing biomass crop on the set-aside land in the context of an aid scheme for renewable raw materials (Bundesministerium für Ernährung, Landwirtschaft und Forsten, 1996; United Kingdom Ministry of Agriculture, Fisheries and Food, 1996b).

The **set-aside option of the agri-environmental regulation** provides an incentive for farmers to withdraw farmland from production for at least 20 years with the goal of achieving environmental improvements, such as the establishment of biotope reserves and nature parks or the protection of water systems. Because of the diversity of environmental conditions and agricultural structures in the Union, Member states are encouraged to implement long-term set-aside on the basis of zonal programmes. They are also called upon to ensure regular monitoring of the programmes and dissemination of the results. The European Commission participates in monitoring the scheme, by collecting information from Member states and performing audit controls. At the national level, agriculture ministries, sometimes in co-operation with environment ministries, are responsible for programme administration (de Putter, 1995).

An annual payment per hectare of up to ECU 600 (US$784), which is co-funded by the Union and the Member State, can be granted to participating farmers. The rate of Community part-financing is 75 per cent in Objective 1 areas, and 50 per cent in other regions. The actual payment rate is determined by Member countries, with approval from the Commission, based on the income loss associated with land diversion, the expected environmental benefits, and the need to provide a sufficient incentive to achieve certain objectives. Countries may raise the payment rate above the ECU 600 level if this is justified by the environmental benefits, but in this case have to finance the difference out of their own budget. Member states may restrict payments to a maximum amount per holding, and differentiate payment rates according to farm size.

Long-term set-aside is offered in all but one of the 14 Member states for which agri-environmental schemes had been approved by mid-1996 (Table EU1). While in some countries, long-term set-aside is confined to arable land, it can also be open to certain non-arable lands as in the case of the United Kingdom. In seven of the countries, annual per-hectare payments for long-term set-aside are specified in terms of lower and upper limits, in four a single payment rate has been fixed, and in

Table EU1.   **Payment rates for set-aside under the agri-environmental regulation[1]**

Country	Minimum payment (ECU/hectare/year)[2]	Maximum payment (ECU/hectare/year)[2]	Fixed payment (ECU/hectare/year)[2]
Belgium		included in other programmes	
Austria	290	604	
Denmark			226
Finland		513	
France			375
Germany	275	595	
Ireland			600
Italy	310	600	
Luxembourg	162	243	
Netherlands		not offered	
Portugal			464
Spain	158	525	
Sweden	260	530	
United Kingdom	146	566	

1.   Greece is not included in the table because no programme has yet been submitted or approved.
2.   1 ECU = 1.31 US$ in 1995.
Sources : de Putter (1995); Finnish government, 1996.

one country a payment ceiling has been set. In Belgium, no separate payment is made because the set-aside option is included in other programmes. The size of the payment rates depends on the management restrictions imposed, the environmental value of the land covered by the programme, and the opportunity cost of farmland and labour. Payment rates can range from ECU 146 (US$191) per hectare for the lowest-paying option in the United Kingdom to the allowed maximum of ECU 600 (US$784) per hectare in Austria, Ireland and Italy (de Putter, 1995).

The objectives of the **forestry scheme** are to improve forest resources, reduce the shortage of wood in the EU, encourage forms of countryside management more compatible with environmental balance, and combat the greenhouse effect. Participating farmers receive a payment towards the cost of afforestation (subject to certain payment ceilings), a degressive five-year payment to cover the costs of maintaining the new woodlands, and an annual compensation payment, which can be granted for up to 20 years, for income losses incurred during the non-productive period of the growing forests. The forestry scheme also offers investment aid for the improvement of existing woodlands.

Member states specify the size of the payments on the basis of actual afforestation expenditure or income foregone, and are obliged to take measures to evaluate and monitor the environmental impacts. Maximum co-funded payments towards the cost of tree planting may vary from ECU 2 415 (US$3 156) to ECU 4 830 (US$6 312) per hectare depending on the tree species. Payments for the maintenance of planted woods can range from ECU 181 (US$237) per hectare to ECU 604 (US$789) per hectare, and annual payments during the non-productive period are limited to ECU 725 (US$947) per hectare. With regard to the improvement of existing woodlands, payments can be granted for planting shelter-belts, constructing forest roads, improving cork oak stands, and creating firebreaks and waterpoints (CR 2080/92). The provisions for programme implementation, including zonal arrangements, Union part-financing, monitoring of compliance and attainment of the objectives are similar to those of the agri-environmental scheme.

Until the 1995/96 marketing year, land diversions under the agri-environmental regulation and the forestry scheme were completely separate from set-aside under the arable aid scheme. Farmers were not allowed to count land taken out under the long-term land diversion schemes towards their annual market set-aside obligation. This arrangement was thought to have created a certain reluctance by farmers to participate in long-term set-aside schemes. As a result, the rules were changed and from 1996/97, arable land taken out of agricultural production under the "accompanying measures" on or after 1 July 1995 can be counted as market set-aside if it satisfies the requirements of the arable aid scheme. The payment rate applicable to this land is that of the agri-environmental scheme. If this rate exceeds the per-hectare payment for market set-aside, only the market-set aside rate is applied. This amendment is expected to increase the incentive for farmers to divert land under the agri-environmental and forestry schemes. It could also lead to a shift of the financing burden from the Union to Member countries, as set-aside under the accompanying measures are co-funded, whereas market set-asides are fully funded by the Union.

Environmental assessment

Discussion of the set-aside rules

Market set-aside under the arable aid scheme is not targeted to environmentally sensitive land. It is implemented on all commercial farms that participate in the scheme and is therefore dispersed over a given region. Also, there are no land selection criteria at the farm level that would target erodible or otherwise environmentally sensitive land. Farmers are likely to choose marginal land over environmentally sensitive land to satisfy their set-aside obligations. Moreover, the short duration of market set-aside severely limits the environmental improvements that can be achieved.

The management rules applying to set-aside land are major determinants of the environmental effects. These rules relate to the establishment of a plant cover, farm input use and the timing of the management activities. Their principle aim is to prevent environmental degradation of diverted land, but they can also contribute to generating environmental benefits for soils, water and wildlife. General rules concerning the minimum period during which the land has to be idled and restrictions on agricultural use of the green cover, if such a cover is established, are set at the Union level, but most of the specific management rules are drawn up and implemented by Member states and can be adapted to suit the countries' environmental and agronomic situation. A comparative study of the implementation of set-aside in several EU countries suggests that significant differences in management rules exist among countries, in particular with respect to the establishment of a cover crop, permitted maintenance practices, the timing of various activities, and the use of chemicals on diverted land (Ansell and Vincent, 1994).

The general rules stipulate that set-aside land be idled at least for the period between 15 January and the end of August. If a green cover requirement is implemented by Member states, the plant cover must not be used for seed production or any other agricultural purpose before the end of August.[14] Grass may be cut but not removed during the idling period. In addition, no marketable crops may be grown after the end of the idling period until mid-January of the following year. However, forage grown on the set-aside plots during this time may be fed to livestock on the farm.

The Management rules drawn up by Member states typically concern establishment of a vegetative cover, either through natural regeneration or through seeding, and restrictions or a ban on the application of fertilisers, animal manure, sewage sludge and chemical plant protection products. To reduce wildlife damage, cutting of grass can be limited to certain periods and use of selected non-residual herbicides for weed control can be allowed. Protection of landscape features and wildlife habitat on set-aside land may also be required. On non-rotational set-aside, a green cover may be required all year round.

The environmental management rules are less stringent if farmers produce non-food crops for industrial use on diverted land. Although Member states are required to apply appropriate environmental measures to ensure protection of the environment, non-food crop production usually remains exempted from many of the environmental restrictions applying to other set-aside land. While non-food crops provide a plant cover, they may also require fertilisation, tillage and treatment with farm chemicals, and can be at odds with the objective of environmental conservation.

The successive modifications made to the market set-aside scheme are likely to lead to a greater concentration of diverted land in marginal areas. The harmonisation of the idling percentage will reinforce the trend from rotational to non-rotational set-aside and encourage diversion of the least productive land of the farm. Similarly, the provision allowing farmers to trade their set-aside liability to other producers within a 20 kilometre radius, could result in a shift of diverted land to the most marginal areas in a region. While set-aside of marginal land might in many situations create additional environmental benefits, this land is not always in need of protection (Council for the Protection of Rural England, 1994). Transfers of set-aside obligations to areas in which specific environmental objectives are pursued have a greater potential for generating positive ecological effects.

The amendment allowing farmers to include long-term set-aside in their annual market set-aside commitments could remove a barrier to enrolling land in environmental set-aside schemes. In the United Kingdom, for instance, the new rules could increase participation in the *Habitat Scheme* and the *Farm Woodland Premium Scheme*, both of which take land out of production for 20 years or more. The new rules could also result in more arable land being converted to extensive grass in *Nitrate Sensitive Areas*, with benefits for water quality (United Kingdom Ministry of Agriculture, Fisheries and Food, 1995). However, the initial reaction by farmers to these new possibilities appears to have been rather reserved, partly because of the long-term commitment required by the environmental land diversion schemes and the uncertainty surrounding the future of market set-aside.

The linkage between market set-aside and the agri-environmental regulation could also induce farmers to place land coming out of short-term set-aside into a permanent scheme, thereby preventing the land from coming back into production. The UK's Habitat Scheme, for instance, includes an option for targeting land diverted under the 1988 five-year set-aside programme with the objective of keeping the land withdrawn for another twenty years (Council for the Protection of Rural England, 1995).

Compared with the market set-aside schemes, which leave little room for environmental targeting, the land diversion measures introduced under the **agri-environmental regulation** and the **forestry scheme** are directly addressed to environmental issues. They also recognise to a greater extent the differences in agricultural and environmental conditions within the Union (Baldock *et al.*, 1993). The agri-environmental regulation, for instance, offers a "menu" of measures from which national governments can select those that are most suitable for the country. Moreover, eligibility and payment criteria can, within certain limits, be determined at the national level. Of the countries that have chosen to offer long-term set-aside as an option for farmers, most have done so by developing a separate programme for land diversion, often in connection with zonal programmes.

Effects on land use

The 1988 **five-year set-aside** scheme was principally intended to stabilise agricultural markets. All Member states had to offer the programme, but farmer participation was voluntary. With the introduction of the arable aid scheme in 1992, the programme was closed to new applications. In 1993/94, over 1.6 million hectares were set aside under five-year contracts, with Italy accounting for more than half of the total. By 1994/95, the area diverted under the programme had declined to 1.3 million hectares (Table EU2). In 1995/96, it had fallen to 846 000 hectares.

Table EU2. **Land set aside under the arable aid scheme**

Thousand hectares[1]

	1993/94	1994/95	1995/96
Voluntary five-year set-aside	1 649	1 296	846
Rotational set-aside	4 614	3 805	2 551
Non-rotational and voluntary set-aside	–	2 197	3 775
Total set-aside	**6 263**	**7 298**	**7 172**
Total base area	48 826	49 033	53 361

1. The numbers refer to EU-12.
Source: Commission of the European Community.

Farmer participation in the five-year scheme remained below expectations and the projected budgetary savings did not materialise because of the large administrative costs of the programme (Brouwer and van Berkum, 1996a). The scheme was voluntary and the only incentive for participation was that provided by the per-hectare payment for idled land. Given the high levels of market price support for grains during that period, many farmers hesitated to join the programme. With the 1992 CAP reforms, set-aside was made a condition for receiving direct payments on the entire arable area, thus increasing the incentive for land diversion.

The implementation of set-aside requirements under the arable aid scheme led to a rapid increase in the total diverted area. In 1993/94, over 4.6 million hectares of cropland were removed from production in the form of **rotational** set-aside. Collectively, almost 6.3 million hectares were idled under the rotational and the five-year schemes (Table EU2). In the following year, the diverted area increased to 7.3 million hectares. The increase was mainly accounted for **by non-rotational** set-aside, which was offered for the first time, and by an estimated 600 000 hectares of **voluntary** set-aside, half of which was signed up in Spain (European Commission, 1995a). The areas under rotational and five-year set-aside both declined.

The largest diverted areas were in France and Germany, followed by Spain and Italy. The composition of the diverted area by type of set-aside differed considerably across countries. In 1993/94, five-year set-aside still accounted for more than half of the diverted land in Italy and the Netherlands, while in Greece, Ireland and Portugal only rotational set-aside was used. The rotational option was also the most important form of set-aside in Belgium, Luxembourg, Spain, France and the United Kingdom. In Denmark and Germany, non-rotational set-aside became the predominant form of land diversion in the first year of its implementation. Voluntary set-aside accounted for a significant proportion of total land diversion only in Spain and France.

With the exception of Italy, national set-aside areas increased in all countries from 1993/94 to 1994/95. In relative terms, the largest increases occurred in Spain (49 per cent), Belgium/Luxembourg (32 per cent), Ireland (29 per cent) and Denmark (26 per cent). The decrease recorded for Italy is explained by a decline in the area diverted under the expiring five-year scheme. The expansion in national set-aside areas was partly due to the decrease in the number of farmers adhering to the "simplified" scheme, and the corresponding increase in the area covered by the regular scheme. A switch from rotational to non-rotational set-aside (which was subject to a higher set-aside percentage), and first-time participation by large farmers who joined the scheme because of higher arable payments, also contributed to the increase in the set-aside area (Brouwer and van Berkum, 1996a).

The total market-set aside area declined in 1996/96 to 7.1 million hectares, mainly as a result of the lower set-aside requirement (Table EU2). Rotational set-aside and land set aside under the five-year scheme decreased by 1.3 million hectares and 450 000 hectares respectively, whereas the non-rotational forms expanded by almost 1.6 million hectares. Non-rotational set-aside accounted for 53 per cent of total market set-aside, as compared to 30 per cent in the previous year.

Total set-aside under the arable aid scheme ranged from just over one per cent of the national base area in Greece to almost 17 per cent in the UK in 1994/95. Germany, France, Italy and Spain had more than 15 per cent of their base area idled, whereas in Belgium, the Netherlands and Portugal, only 6 per cent was set aside. This variation is largely explained by the farm size distribution, which leads to a high proportion of farmers being exempt from the set-aside obligation in Greece, Portugal and the Netherlands, and by voluntary set-aside, which accounts for Italy's high national percentage.

An estimated 686 000 hectares of diverted land were used for the production of non-food crops in 1994/95, as compared to 264 000 hectares the year before. France and Germany accounted for 60 per cent of this area. About 600 000 hectares of the non-food crop area were sown to rapeseed and sunflower seed. In 1995/96, non-food oilseed production on set-aside land was estimated to have attained 970 000 hectares (Bundeministerium für Ernährung, Landwirtschaft und Forsten, 1996). Most of the rapeseed is used for the production of biofuels.

Total production of oilseeds on set-aside land is limited by the Blair House Accord between the United States and the European Union, which stipulates that the by-products of oilseed production for non-food use must not exceed 1 million tonnes of soya bean meal equivalent. Under current conditions, one hectare of oilseeds corresponds to roughly one tonne of soya bean meal equivalent (European Commission, 1995b).

Long-term set-aside under the agri-environmental regulation is mainly being implemented to improve water quality, restore the natural flora and fauna, reduce soil erosion, and prevent natural disasters. National or sub-national schemes can be fairly specific in their objectives. In Hesse (Germany), 20-year set-aside is targeted to riverbanks and meadowlands; in Luxembourg, the primary goal is habitat protection; in the Alsace region of France, the predominant objective is water quality improvement; and a Portuguese programme is aimed at protecting volcanic lakes in the Azores.

Long-term environmental set-aside is often integrated with zonal programmes, which apply to specific areas and have a limited possibility for land enrolment (de Putter, 1995). Due to the recent implementation date of agri-environmental programmes in many countries, systematic enrolment figures for long-term set-aside are not yet available. In Denmark, for instance, the programme was offered for the first time in 1995/96 (Schou, 1995).

The European Commission has approved national and regional programmes under the **forestry scheme** that envisage afforestation, maintenance and improvement of forests on around 930 000 hectares of land by 1997 (Communication by the European Commission).[15] By far the largest area (more than 300 000 hectares) has been committed in Spain, followed by Italy, Portugal and the United Kingdom. According to the Commission, afforestation under the forestry scheme could become an attractive alternative for using part of the 10 to 15 million hectares of land that might be removed from traditional agricultural production in the near future (European Commission, 1994).

Effects on the environment

Most of the land diverted under the 1988 **five-year set-aside** scheme was of poor quality. It was either left fallow, planted with trees or used for non-agricultural purposes. The programme resulted in the creation of extended areas of grassland, a cut-back in farm chemicals applications, and a reductions in the use of animal manure on diverted land (Hawke et al., 1993, Hawke and Kovaleva, 1994).

The environmental effects of **market set-aside** are site-specific and depend on natural factors and the management practices applied. In many cases, a risk of environmental degradation is created on the idled land. The types of risk depend on the climate, the soil type and the farming system, and include wind erosion during the dry season, water erosion on steep slopes during the winter, nutrient leaching on soils that are saturated with nutrients or have a low retention capacity, and invasion of weeds and pests. However, experience with market set-aside suggests that negative environmental degradations on idled land can generally be prevented if appropriate management rules are established and observed by farmers. Similarly, the positive effects of set-aside depend on the way set-aside is managed, but also on the type of land diverted and the length of the set-aside period.

There are indications that fallowing land may have led to improvements with respect to soil productivity, water quality and biodiversity, particularly in regions which are predominantly used for intensive cropping. Selection of appropriate cover plants may have enriched the soils with *organic matter* and *nutrients*, especially nitrogen. The benefits from a seeded green cover are in many respects superior to those of natural regeneration. The quantity of nitrogen contained in the vegetative matter at the end of a one-year idling period, for instance, has been measured to be 25 times higher if certain leguminous plants and grass mixtures are grown than under natural regeneration (Godden et al., 1994). On the other hand, the effect on biodiversity can be more beneficial under natural regeneration. The positive effects of green fallow on soil productivity might increase as a result of the recent shift from rotational to a non-rotational set-aside.

Since most of the diverted land does not receive any fertilisation, set-aside may have reduced overall *fertiliser* applications in production. This impact is likely to diminish with the increase in non-rotational set-aside. A study of nitrogen use in north-western England suggests that with a uniform diversion percentage, non-rotational set-aside is likely to reduce total nitrogen use by less than rotational set-aside (Rygnestad and Fraser, 1996). Non-rotational set-aside permits to idle land of poorer quality and leave the more fertile land, with its associated higher nitrogen application rates, permanently in production.

With respect to *water quality*, rotational set-aside can have harmful long-term consequences. Leaving land fallow in high-rainfall cropping systems can lead to leaching of fertilisers and pesticides into the groundwater. The risk of nutrient leaching is particularly high at the beginning of the set-aside period when nitrate uptake by plants declines as compared to continued intensive farming. In cases where set-aside leads to an accumulation of vegetative matter in soils that are already rich in nutrients, there can also be a substantial risk of nitrate emissions from the ground when the land is returned to production (Williamson, 1993). On the other hand, studies in the UK have shown that the risk of nitrate leaching on rotational set-aside need not be greater than under normal cropping if the land is properly managed. The requirement to have a vegetative cover established by mid-January is designed to keep the risk of nitrate leaching low.

Soil erosion is a potential problem on set-aside land, especially when the green cover is not established fast enough after the harvest and the land is exposed to wind and rainfall in the winter. Soil degradation in turn can cause a decline in the soil pH-value and increase the risk of a release of *heavy metals* into the ground-water in areas where such metals have accumulated in the soils. However, there is little evidence so far of changes in the soil pH due to set-aside, and where such changes occur farmers are permitted to apply lime to the diverted land.

Set-aside can have a positive or negative effect on *weeds*. If weeds are allowed to grow on idled land but are destroyed before they can flower, the weed bank in the soil may be reduced. However, grass cuttings in spring before the weeds come into seed can cause damage to wildlife, especially nesting birds, and thus create a conflict between environmental objectives. Where the set-aside land is managed inadequately, persistent weeds can spread and create a need for increased herbicide applications when the land is returned to production, but also on adjacent fields.

Natural regeneration of the plant cover can benefit *wildlife habitat* and *plant diversity* if the land has a short history of intensive cropping and the seed supply in the soil can still generate a diverse local flora (Hawke and Kovaleva, 1994). In other areas, a selected plant cover for birds or game may be more beneficial. Set-aside has been shown to provide valuable winter feeding and nesting habitat for farmland birds. Set-aside after a cereal crop can provide an over-winter cover of stubble which, in combination with a thin layer of natural vegetation, has proven to be very beneficial for certain birds. However, such benefits are being eroded by the reductions in set-aside rates and the introduction of non-food crops to set-aside land (Rayment, 1995).

On one-year set-aside, the vegetative cover will not develop beyond the initial stages of the natural cycle. By comparison, non-rotational set-aside allows for the restoration of a greater variety of habitats and a build-up of local plant and animal populations. But even on non-rotational set-aside can it take a long time until a richly diversified flora develops, and the newly created habitat may never equal the ecological balance that had prevailed before the advent of intensive farm production.

In general, the positive effects of set-aside on wildlife and biodiversity will increase with the size of the diverted plots. Trade in set-aside obligations in combination with multiannual land diversion can lead to local concentrations of idled land, thereby increasing the potential for environmental improvements (Sebillote *et al.*, 1993). On the other hand, relatively narrow strips of land can support a rich biodiversity and be of strategic importance for resource conservation, especially in field margins and riparian zones where different ecological systems meet. Placed under a perennial grass cover, set-aside in field margins, riparian areas and low wetlands can have significant benefits for water quality. The minimum size restriction for diverted land parcels (at least 20 metres wide) has prevented farmers from including more of these lands in their mandatory set-aside (United Kingdom Ministry of Agriculture, Fisheries and Food, 1996a).

The ecological value of *forest plantings* depends significantly on the type of trees planted, and will be higher for mixed forests than for fast-growing energy forests with only one or a few tree species. Market set-asides offer only limited possibilities for improvement, as only forests with a very short harvest cycle can be grown. On the other hand, fast-growing forests need little or no herbicide and nutrient applications, and can be grown on sandy soils (Wagner, 1995). Fast-growing tree species used on set-aside land include willow and poplar, but also birch, alder, and robinia.

Another potential benefit of land diversion is *greenhouse-gas* abatement. Reductions in the CO_2 level can be achieved through afforestation of set-aside land (expanding the CO_2 sink), development of renewable energy resources to replace fossil fuels, and a shift to less energy-intensive farming practices. There is some indication that especially biomass production could offer possibilities for greenhouse-gas reductions (Hawke and Kovaleva, 1994).

With respect to *landscape*, the effects of set-aside are perceived to be overwhelmingly negative, especially when the land is left to natural regeneration (Council for the Protection of Rural England, 1994, 1995). Afforestation on open land can enhance the landscape, but if trees are planted in areas with an already high percentage of forests, there is a danger that further plantings might reduce the recreational value of the countryside. Studies on the relationship between forest cover and tourism indicate that there is no negative relationship below forest coverage rates of 80 per cent (Wagner, 1995).

Production of *non-food crops* on set aside land does in principle not differ from conventional production. The environmental effects will largely depend on the crop planted and the production methods used. Rapeseed, for instance, is a crop whose cultivation requires substantial amounts of nitrogen and phosphorus, and often fertilisation (Wagner, 1995).

Many of the environmental benefits created on diverted land, especially those related to the creation of wildlife habitats and biotopes, are *temporary*. They are lost when the land is brought back into production. Recent attempts to prolong the effective duration of set-aside by targeting the same land through successive schemes, or encourage enrolment of land coming out of market set-aside in an environmental scheme could lead to improvements in this respect.

Set-aside was once an integral part of traditional farming practices in Europe, yet its large-scale policy-driven reappearance poses a number of agronomic and economic problems. Due to a lack of recent experience with set-aside as part of the crop rotation, farmers are often unsure of the environmental consequences of alternative management choices. Although management rules have been established by Member countries, additional research into the most appropriate ways of incorporating set-aside into crop rotations may be necessary to achieve further improvements (Institut National de la Recherche Agronomique, 1995).

The long-term land diversion schemes under the **agri-environmental regulation** and the **forestry scheme** have only recently been implemented and it is still too early to analyse their impacts on the environment.

Opinions differ as to whether land diversion schemes have led to a more intensive use of the *land remaining in production*. Some analysts believe that there is little evidence of farmers increasing intermediate input use on land that has not been diverted (Hodge, 1992), whereas others claim that set-aside encourages intensive farming on the remaining land (Council for the Protection of Rural England, 1995). In principle, a reduction in the land available for cropping could lead to a substitution of fertilisers and farm chemicals for land in the production process, although in many cases the price relationships that determine the optimal input mix may not change and there will be little incentive for input substitution. Moreover, farmers with a fixed labour base and no off-farm employment opportunities might shift to more labour-intensive farming practices and decrease intermediate input use. While it is still unclear if set-aside has affected fertiliser and chemicals use on the land remaining in production, the issue is potentially important in connection with land diversion schemes.[16]

Budgetary expenditures

Farmers received an estimated ECU 1.7 (US$2) billion in 1993 and ECU 2.4 (US$2.9) billion in 1994 from the EU budget for idling land under the *arable aid scheme*, including five-year voluntary set-aside. In 1994, expenditures on the five-year scheme accounted for 25 per cent of the total; in 1995, the corresponding figure had declined to 10 per cent. The largest payment recipients were France, Germany and Spain, which together accounted for nearly three-quarters of the total budgetary expenditure in 1995.

With respect to the *agri-environmental* programme, expenditure figures for the individual elements of the scheme, including *long-term set-aside*, were not available by mid-1996. Union expenditures on *all measures* implemented under CR 2078/92 were around ECU 123 (US$144) million in 1993, ECU 231 (US$275) million in 1994, and ECU 520 (US$680) million in 1995. Over the period 1993-97, the total budgetary cost of agri-environmental programmes developed by Member states and approved by mid-1995 is estimated to be around ECU 6.5 (US$8.5) billion, with a cost-sharing contribution by the Union of ECU 3.7 (US$4.8) billion (de Putter, 1995).

Union expenditures on the *forestry scheme* were around ECU 7 (US$8.2) million for 1993, ECU 83 (US$99) million for 1994, and ECU 200 (US$261) million for 1995. The total budgetary cost of the scheme over the period 1993-97 is estimated to be in the area of ECU 2 (US$2.6) billion, of which roughly ECU 1.3 (US$1.7) billion are financed by the Union (Communication by the European Commission). Spain and Italy account for more than half of the total expenditure. Because of differences in payment rates across countries, the budgetary cost for a country is not necessarily proportional to the area signed up under the scheme.

Overall, government payments under land diversion schemes account for only a small proportion of total agricultural support in the EU as measured by the Producer Subsidy Equivalent. The total PSE has been estimated at ECU 67 (US$88) billion in recent years (OECD, 1996). However, budgetary expenditures for land diversion schemes have been among the fastest growing types of direct payments to farmers.

Effects on grain production and resource use

Given the small-farmer exemption from the set-aside constraint, it was clear from the beginning that the supply effect of the arable aid scheme would be considerably below the announced set-aside percentage. Successive modifications to the scheme have further weakened its effectiveness as a means of supply control. Estimates suggest that the **area planted to cereals** in the EU-12 in the 1996/97 marketing year was only 0.2 per cent below the pre-reform (1992/93) area. Although cereals plantings had dropped by 8.3 per cent in 1993/94, this decline was followed by increases of 0.9 per cent, 2.9 per cent and 4.8 per cent in the subsequent years. In Germany and the Netherlands, the area sown to cereals rose even during the first four years of the scheme. The countries that recorded the biggest declines in cereals plantings are Greece (−22 per cent) and Spain (−8 per cent) (Agra Europe, 26 April 1996). The market set-aside requirement applies also to oilseeds and protein crops, and a full analysis of the production effects of set-aside would have to include developments in the areas of these crops, as well as shifts between oilseeds, protein crops and cereals.

The diminishing effect of set-aside on the cereals area is explained by a combination of factors, including the successive reductions in the mandatory set-aside requirement and the decision to allow 85 per cent of land in voluntary set-aside to be counted against penalty set-aside. Reductions in voluntary set-aside due to favourable world market prices for grains, and a switch from other crops to cereals, may also have contributed to this development. The recent decision to permit inclusion of long-term environmental set-aside in the set-aside commitment will further weaken the effect of the scheme on cereals plantings.

Other changes, such as the introduction of trade in set-aside liabilities and the shift from rotational to non-rotational set-aside, are likely to have increased the concentration of set-aside on less productive land, and may also have diminished the supply effect of the scheme. However, since the shift to non-rotational set-aside has also reduced the positive effect of land rotation on soil fertility, its overall implication for grain production could have been rather limited.

Regarding relative **resource costs,** there are indications that land diversion may be a less cost-efficient means of supply control than extensification. Field experiments in France suggest that by reducing intermediate input use on the cereals area, extensification measures may achieve supply reductions with greater savings in resources than land diversion (Vercherand, 1996). The larger reductions in intermediate input use associated with extensification may also lead to lower environmental risk compared with set-aside. However, a set-aside policy can have environmental benefits that can not be obtained by extensification, especially in the case of longer-term set-aside.

Relationship with other policies in agriculture

The benefits and costs of the EU's land diversion measures depend on the other elements of the Common Agricultural Policy. The price guarantees provided by the CAP have been one of the main driving forces behind the increase in agricultural production in the EU during the past three decades, contributing to the greater use of fertilisers, plant protection products and feed concentrates. Technical progress has also played a role in this development, although technical advances may in turn have been stimulated by support policies. By providing incentives for land consolidation and increases in specialisation, the CAP may also have altered the cultural landscape and increased the pressure on semi-natural habitats (Brouwer and van Berkum, 1996b). By the early 1990s, for instance, the agricultural sector was recognised as one of the main sources of water pollution (Commission of the European Communities, 1992).

The **1992 CAP reforms** initiated a period of lower price support and greater reliance on direct payments. They also marked the beginning of a widespread use of conditionality to safeguard the environment, and the introduction of specific agri-environmental measures. The most notable changes took place in the **arable sector.** Intervention prices for cereals were reduced by one-third over three years and price support for oilseeds and protein crops was eliminated. Lower prices for cereals and oilseeds may lead to different environmental effects across Member states, depending on farming systems and natural conditions. In general, they may induce farmers to reduce their use of fertilisers and plant protection products, and shift production on irrigated land to crops with lower water needs. However, some land could be moved out of cereals production and into crops that require higher applications of chemicals, such as fruits and vegetables, or potatoes.

Regarding the **agri-environmental regulation,** a number of measures were introduced in addition to long-term set-aside. The regulation requires EU Member states to develop agri-environmental programmes and submit these to the Commission for approval. For the countries of EU-12, the proposals had to be submitted by the end of 1993. In general, the programmes should be for a duration of five years, except for long-term set-aside, which is for a period of at least 20 years. By mid-1995, the Commission counted 93 co-financed programmes at the national and regional levels, including zonal programmes.[17] Some Member states submitted only programmes that apply to the entire national territory, others developed regional programmes in addition to national programmes, and both often include zonal arrangements. Ten programmes were implemented in 1993, 44 in 1994 and 33 in 1995. Preliminary projections of participation rates in the agri-environmental scheme range from 3 per cent of the agricultural area for the Netherlands to 91 per cent for Austria (de Putter, 1995).

The measures implemented under the agri-environmental regulation provide incentives for farmers to reduce fertiliser and chemicals use, extensify livestock production, maintain the landscape, adopt organic farming practices, preserve local animal breeds and plant species, prevent environmental degradation of abandoned farm- and woodland, provide public access to land for leisure activities, and convert arable land to extensive grassland. The measures are not restricted to producers who *switch* to one of these activities, but can also be claimed by farmers for *continuing* such an activity. In most cases, the contracts are for a duration of five years. Implementation of the measures is based on proposals developed by national and regional authorities in Member states. By mid-1995, for instance, *measures for the conversion of arable land to grassland* had been developed by seven countries and approved by the Commission (de Putter, 1995).

The size of the annual per-hectare payment depends on the environmental restrictions accepted by farmers. In Germany, for instance, producers may receive DM 150 (US$105) per hectare if they dispense with herbicides in crop production,

DM 150 (US$105) if they farm without commercial fertilisers, and DM 250 (US$174) if they use neither commercial fertilisers nor plant protection products. These payment rates apply to farmers who do not meet the requirements before signing up for the programme. Those who already conform with the conditions at the time of joining the scheme receive a 20 per cent smaller payment. A similar "menu" of choices is offered for grass-based production activities: payment rates can vary depending on whether farmers lower stocking densities by reducing the herd size, increasing the forage area through land purchases, or converting arable land to extensive pasture. Use of commercial fertilisers, plant protection products and irrigation is prohibited (Bundesministerium für Ernährung, Landwirtschaft und Forsten, 1996).

The effectiveness of the agri-environmental regulation will depend on many factors, including the degree of environmental targeting, the provision of information to farmers, monitoring of progress, and integration with other policies. Implementation of the measures requires major administrative expertise by regional authorities. The lack of organisational capacity and experience may limit the potential of the programme, especially in countries that have never before implemented schemes which pay farmers for undertaking specific environmental practices (Brouwer and van Berkum, 1996b). Moreover, countries with a large variety of farming and ecological systems may face a scarcity of scientific and technical information regarding the behaviour of different ecosystems and the contribution of alternative agricultural practices to nature conservation.

Poor integration with other CAP measures (commodity regimes, market set-aside, the forestry scheme, and rural development programmes) can also pose problems. There are examples where EU support for irrigation is in contradiction with the environmental objectives pursued by other measures (Palomo, 1994; Bermejo, 1994), or where parcels idled under the arable aid scheme are located next to land that is irrigated and intensively used (Vercherand, 1996). Monitoring towards the achievement of environmental objectives is currently also insufficient (Brouwer and van Berkum, 1996b).

The **Environmentally Sensitive Area** (ESA) scheme is an agri-environmental measure that was introduced before the 1992 CAP reforms. ESAs are local or regional arrangements that can be implemented by Member countries as part of a 1985 European Council regulation aimed at improving the efficiency of agricultural structures (CR 797/85). This regulation authorises Member states to grant aid to farmers in "environmentally sensitive areas" who agree to adopt farm management practices that preserve or improve the environment. Although initially not co-funded by the Union, ESAs became eligible for Union support in 1987, and were integrated into the agri-environmental scheme in 1992. Land entered into an ESA may under certain circumstances qualify as market set-aside.

ESAs have been implemented on a large scale in the United Kingdom in areas where wildlife, landscape and recreational values are threatened by agricultural activity. The schemes are voluntary and participating farmers are paid to continue farming in a environmentally friendly manner, which is typically low-input and livestock based. The agreements signed by producers are for a duration of ten years. They may require farmers to reduce fertiliser use and livestock densities, and include prohibitions on the use of herbicides, pesticides and the installation of new drainage or fencing. Producers may also be required to maintain hedges, ditches, woods, walls and barns, and protect historic features. In some cases, farmers may have to reinstate ploughed-up grassland. In return, they receive annual per-hectare payments, which can vary considerably in size depending on the area (Baldock *et al.*, 1990).

EU environmental policy can also give rise to zonal arrangements in agriculture. Under the 1991 European Nitrate Directive, preventive measures have to be taken by Member states in areas where, due to agricultural activity, nitrates in surface water bodies, especially those serving as drinking water sources, or in groundwater, exceed or risk to exceed 50 mg per litre, or where there is evidence or a risk of eutrophication. In 1996, the British government decided to designate 68 **Nitrate Vulnerable Zones** in England and Wales. Farmers in these zones will be obliged to cut back on the amount of animal manure applied to their land, which could have implications for animal stocking densities and manure storage. At the same time, affected farmers will be able to benefit from a grant of 25 per cent towards the cost of improving manure storage and handling facilities.

Concluding remarks

Land diversion programmes have become an important element of agricultural policy in the European Union. While the early programmes were mainly designed to stabilise production, the current schemes consist of two basic types: annual and medium-term set-aside for supply control, and long-term land diversion for environmental purposes. Characteristic features of the current schemes include a comprehensive set of management rules for diverted land, a considerable scope for adaptation of the programmes to regional and local environmental needs and conditions, and a decentralised implementation process. The large-scale introduction of set-aside in the EU creates a number of uncertainties, because there is a great variety of ways in which diverted land can be managed and the implications for the environment of a particular management choice are not always clear. Identification of the most suitable set-aside practices will involve a learning process for farmers and programme administrators.

Market set-aside under the arable aid scheme is primarily a supply control measure. Idled land has to be managed in ways appropriate to prevent environmental degradation and generate, where possible, benefits through improvements in soil and water quality, creation of bird and wildlife habitat, and protection of landscape features. Economic use of set-aside land is prohibited except for the production of certain non-food crops. The rules for set-aside have been modified several times, resulting, among others, in the introduction of limited tradability and the creation of a link between market set-aside and long-term environmental set-side.

Rotational set-aside is the predominant form of set-aside in many EU countries, although a shift to the non-rotational form has been observed in recent years, and the decision to apply the same set-aside percentage to both forms in the 1996/97 marketing year can be expected to further reduce the importance of rotational set-side. The environmental benefits of rotational set-aside are limited by the short duration of the idling period, yet improvements in soil organic matter, nutrient balance, and plant and wildlife populations can be achieved through careful management of the vegetative cover. Among the problems that can be encountered on annual set-aside is an increased risk of erosion if the plant cover is established too late or incompletely, nutrient leaching into the groundwater and a proliferation of persistent weeds, which can create a need for increased weed control when the land is prepared for the next cropping season. However, environmental degradations on annual set-aside can largely be prevented if proper management rules are established and enforced.

The *non-rotational* form can have more favourable impacts on the environment and landscape than the rotational scheme. If land is set aside for several years in a row, a greater diversity of plant species is likely to develop, which will subsequently support a greater variety of wildlife. A multiannual plant cover reduces the risk of erosion and can improve soil structure and nutrient balance. However, even with non-rotational set-aside, many of the environmental improvements achieved on idled land are temporary, as the land is eventually returned to production. Recent attempts to encourage land coming out of multiannual set-aside to be enrolled under an agri-environmental scheme could help secure the environmental benefits that have accumulated during the set-aside period.

The impacts of *non-food crop production* on the environment are not much different from those of normal cropping. Since diverted land used for non-food crops is not subject to the management restrictions that normally apply to set-aside, potential benefits are limited to an increase in cropping diversity and, in the case of biomass production, medium-term habitat creation and landscape effects.

The impact of market set-aside on *cereals production* has decreased since the introduction of the arable aid scheme. In the 1996/97 marketing year, the area sown to cereals was only slightly smaller than in the pre-reform area. Successive reductions in the land diversion percentage, the introduction of trade in set-aside liabilities, the possibility to deduct part of voluntary set-aside from penalty set-aside, the return of land from the expiring five-year scheme to production, and a reduction in voluntary land diversion due to favourable world market prices for grains have contributed to this development. The shift from rotational to non-rotational set-aside could also have been a contributing factor.

Long-term set-aside under the agri-environmental and forestry schemes provide an opportunity for a better targeting of environmental objectives and a high degree of regional flexibility. Long-term set-aside is currently offered in 12 EU Member states, and almost one million hectares of land have been committed to afforestation under the forestry scheme. Although it is too early to judge the implications of these measures for environment and landscape, the benefits are likely to be greatest where lands of high environmental value are attracted into the programme, and where these lands are managed in ways that are adapted to local ecosystems.

The payments for long-term set-aside are related to income foregone, but no attempt is being made to encourage low-cost ways of providing the environmental services, such as by using a competitive bidding process for land enrolment. Moreover, there are no provisions to estimate, for each land parcel offered by farmers, the environmental benefits that would be expected from placing the land into the programme, and use these estimates as a criterion for land acceptance. Several other factors could also reduce the effectiveness of long-term set-aside, including a lack of data on land of high ecological value and insufficient knowledge of the contribution of alternative land uses to nature conservation. Care will also have to be taken to ensure adequate integration with the other measures of the CAP, and implementation of appropriate monitoring and evaluation systems.

The EU's land diversion schemes reflect the increasing recognition of environmental issues in the CAP. The **1992 reforms of the CAP** have led to a reduction in price support, the attachment of environmental conditions to direct payments, and the introduction of programmes specifically aimed at safeguarding the environment. Lower institutional prices for *arable crops* may have induced farmers to use less fertilisers, plant protection products and irrigation water, thus contributing to the decline in intermediate input use witnessed in many EU countries.

The *agri-environmental regulation* provides farmers with a series of incentives to extensify production, maintain the landscape and preserve biodiversity. Implementation is based on national and regional plans and offers opportunities for flexible targeting and adjustment to local conditions. Yet, it also requires major technical and administrative expertise on the part of regional and local authorities. The lack of organisational capacity and experience could limit the potential of the programme, especially in countries that have never operated similar schemes before.

The *budgetary expenditures* on land diversion schemes are still relatively small when compared to total agricultural support, although they have been increasing rapidly. While the budgetary cost of short-term set-side is linked to world market developments for arable crops, expenditures on environmental set-aside and other measures of the agri-environmental and forestry schemes are based on medium- to long-term commitments. A full assessment of these expenditures would require better knowledge of the actual environmental improvements obtained from these measures. The development and implementation of appropriate environmental monitoring systems, including agri-environmental indicators, could help ensure that the taxpayer funds expended on these programmes are guided to their most effective use.

5. JAPAN'S RICE PADDY FIELD DIVERSION PROGRAMMES

Background

Japan is a mountainous country with over 67 per cent of its area covered by forests. Only 14 per cent of the land area is used for agricultural production, and a similar proportion is reserved for national or prefectural parks. Around 37 per cent of Japanese agricultural production takes place in mountainous areas, accounting for about 42 per cent of the total agricultural land. Japan's high rate of precipitation, in combination with steep river gradients and mountain slopes, places the country under a permanent risk of flooding and water erosion.

In 1994, around 2.2 million hectares or 44 per cent of the agriculturally used area was cropped to rice, with some 2.83 million farm households engaged in rice production. The average rice cropping area per farm household is slightly greater than three-quarters of a hectare. Almost all of rice production takes place in paddy fields. Roughly 40 per cent of the paddies are on terraced fields in hilly areas, whereas the other 60 per cent are situated on relatively flat lowlands, often alluvial plains. Paddy fields are surrounded by artificial ridges or farm roads which can temporarily contain the water flow and slow the discharge of water into streams and rivers. Paddies constitute an important tool for water management, flood prevention, groundwater infiltration, erosion control and, in the hills and mountains, landslide protection.

Not all paddy fields are used for rice production. Of the total area covered by paddies in 1994 (2.7 million hectares), over 500 000 hectares were used for purposes other than rice production. The ridges surrounding paddies, which account for roughly six per cent of the total paddy field area, also reduce the area that is available for producing rice.

With the appearance of rice surpluses in Japan in the early 1970s and the government's choice of land diversion as a means of production control, concern has been created about the environmental effects of withdrawing paddy fields from rice production.

The Programme

The Japanese government has operated programmes to divert land from rice production since 1971 (see Table J1). The total area diverted has varied considerably from year to year, ranging from one-quarter of a million hectares in 1975 to over three-quarters of a million in the early 1990s. In most years since 1971, more than 500 000 hectares were withdrawn from rice production. The **Paddy Field Farming Revitalisation Programme** (PFFRP), which was in place from 1993 to 1995, and it successor, the **New Production Adjustment Promotion Programme** (NPAPP), which covers the 1996-98 period, are the two latest programmes in this series.

Tableau J1. **Japanese rice paddy field diversion programmes**

Land diversion programmes (selected years)	Area diverted (1 000 hectares)	Total subsidy payments [billion yen (billion US$)]	
1971	541	171.1	(0.51)
1975	264	92.8	(0.31)
1980	585	357.8	(1.58)
1985	594	222.5	(0.93)
1990	849	159.8	(1.10)
1991	852	156.7	(1.17)
1992	751	130.0	(1.03)
1993	713	90.2	(0.81)
1994	588	63.3	(0.62)
1995	659	80.7[2]	(0.86)
1996	787[1]	55.9[2]	(0.59)

1. National target value for 1996;
2. Budget allocation.
Source: Japanese government, 1996.

The paddy field diversion programmes have a predominant objective: to reduce rice surpluses in the face of increasing productivity and declining demand. Having primarily been conceived as a supply control measure, they had originally no specific environmental objectives. However, since 1978 environmental provisions have been gradually incorporated into successive programmes, and more recently, paddy field diversion payments have been classified as an environmental measure within the Uruguay Round Agreement on Agriculture, thus not being subject to support reduction commitments. In the NPAPP, environmental concerns are formally recognised as a policy objective.

The environmental provisions associated with the paddy field diversion programmes are preventive in nature and are intended to counteract potentially negative environmental effects of withdrawing, on a temporary or permanent basis, paddies from rice production.

The government has identified several types of alternative paddy field uses that could contribute to this objective, and farmers who limit themselves to these alternatives become eligible for acreage subsidies. In practice, the large majority of farmers opt for a subsidised use of diverted land. Under the NPAPP, farmers are also urged to make efforts to reduce the environmental pressures created by agricultural activities on diverted land, and to implement appropriate measures to this effect under the instruction of municipalities and farm co-operatives.

The two main subsidised land use options are alternative cropping and "managed" idling of paddy fields (Table J2). In 1993, paddy fields planted to alternative crops accounted for 87 per cent of diverted land receiving government payments. The main crops were fodder crops, wheat, soya beans, feed grains and vegetables (Table J3). Idled paddies that were kept in their original state accounted for nine per cent of paid land diversion. In both cases, the productive characteristics of paddy fields are maintained so that the paddies can be returned to rice production if necessary.

Less use was made of the third subsidised land use option available to farmers in 1993, the conversion of paddy fields to forests, orchards or fish ponds (Table J2). Under this option, some of the traditional environmental effects of paddy fields are

Table J2. **Land use on diverted rice paddy fields in 1993**

	Environmental conservation function	Area (hectares)
Types of land use eligible for government payments in 1993		
Planting alternative crops in paddy fields	Prevention of soil erosion and landslides; prevention of floods	428 000
Maintaining idled paddy fields (maintenance of ridges; weed control)	Conservation of water resources; prevention of soil erosion and landslides; prevention of floods	46 000
Converting paddy fields to other uses (forests, orchards, fish breeding ponds)	Conservation of water resources; prevention of floods; prevention of soil erosion and landslides	13 000
Sub-total		487 000
Types of land use not eligible for government payments in 1993		220 000

Source : Japanese government, 1995.

Table J3. **The main alternative paddy field crops, 1993**

Crop	Area in hectares	Crop	Area in hectares
Pasture	67 000	Sorghum	12 000
Wheat	66 000	Red bean	12 000
Soya bean	41 000	Buckwheat	11 000
Soil-improving crops	29 000	Eggplant	10 000
Corn	15 000	Flowers	10 000

Source: Japanese government, 1995.

lost, yet the permitted alternatives, including forests and fruit trees, also have positive effects with respect to erosion control and flood and landslide prevention. In 1995, a further subsidised land use type was added to the list: maintaining rice production capacity by keeping paddy fields flooded with water.

In addition to choosing a recognised type of land use, farmers have to carry out activities relating to the maintenance of ridges and weed control to establish eligibility for subsidies. Municipalities are charged with ensuring that the conditions for subsidy payments are met. The payments are made on an annual basis. For permanent paddy field conversion the payment is granted for a fixed number of years.

Around 31 per cent of the paddy fields diverted from rice production in 1993 were not eligible for subsidies (see Table J2). This area consisted of land used for processing rice production, which was not a subsidised land use option but counted towards the set-aside obligation, and of land that had been permanently withdrawn from rice production in the past and which was no longer eligible for government payments because the fixed payment period had ended before 1993. As from 1996, with the implementation of the NPAPP, production of processing rice and permanent conversion of paddy fields is no longer counted towards the set-aside obligation.

The list of plants that can be grown on diverted paddy fields while maintaining eligibility for subsidies, has grown over the years. The successive additions reflect an evolution in the attention paid to the environmental effects of land diversion. In particular, the inclusion of soil-improving crops on diverted land in 1987, and of landscape-enhancing crops in 1990, constitute efforts to strengthen the environmental conservation element of the programmes.

It is important to note that prior to implementation of the NPAPP, the environmental provisions of the paddy field diversion programmes were fairly limited in scope and did not address many of the environmental pressures that have been created by the increasing intensification of agricultural production in recent decades. Japanese agricultural policy, based on high levels of market price support by OECD standards, is likely to have contributed to this development. Negative environmental impacts of agriculture in Japan are associated with inappropriate use of fertilisers and pesticides, pollution from intensive livestock production and damage to natural habitats (Parris and Melanie, 1993). Such environmental considerations did not enter into the selection of land that was withdrawn from rice production, across regions and on a farm, nor were there environmental compliance restrictions influencing the choice of farming practices on diverted land, including the amount and frequency of fertiliser and pesticide applications, beyond the standards that applied to farmland in general.

The NPAPP has made environmental management of diverted land a requirement for payment eligibility, thus providing a basis for addressing input-related environmental problems. In principle, this could involve application of appropriate conservation measures to the more environmentally sensitive land. There are no binding rules at the national level, but the Japanese government has urged municipalities and farm co-operatives to implement such measures.

The national land diversion target is determined every year after consultation between the government and producer groups based on an assessment of the market situation for rice, and allocated across prefectures, municipalities and individual rice growers. In particular, the current programme specifies that adjustments in the national target area should be aimed at reducing the carryover stocks of domestic rice to the yearly amount considered necessary for stockpiling purposes (approximately 1.5 million tonnes).

Farmers can decide which part of their paddy fields to divert to satisfy individual land diversion requirements. The same plots of land may be set aside in successive years. Until 1995, farmers who diverted only part of the individual target area in a given year were penalised by having to set aside a larger area in the following year. Those who refused to participate in the diversion programme and continued to produce rice on their part of the national target area, risked exclusion from other government subsidy schemes. With the expiration of the PFFRP in 1995, these penalties were eliminated. Instead, the NPAPP provides farmers with a financial incentive to comply with the set-aside obligations: if the farmers in an area comply, the government subsidises the mutual relief fund administered by the local farmers' group.

Farmers who participate in the supply reduction programme receive an annual per-hectare payment for their diverted land. The market value of the alternative crop is taken into account in the calculation of the payment rate. In the case of annual cropping on diverted land, the base payment rate under the NPAPP has been set at ¥ 7 000 (US$74) per 10 ares. This payment is increased if farmers take measures that encourage a more efficient use of water and land, or lead to a better integration of crop and livestock operations. Higher payments are granted if, for instance, rice farmers co-operate in order to take adjacent land parcels out of production to facilitate land management, and also if a feed crop is grown on the diverted land under a contractual arrangement with a livestock operation.

The annual payments made for long-term or permanent conversion of paddy fields are higher than the base rate for annual crops, although they are limited to a fixed number of years. In the case of conversion to fruit trees, for instance, payments are made for four years. In the case of conversion to forests, a one-time payment is made. In 1996, the lowest subsidy for paddy field diversion was ¥ 4 000 (US$43) per 10 ares, and the maximum subsidy ¥ 50 000 (US$531) per 10 ares.

Budgetary expenditures on subsidy payments peaked at ¥ 358 (US$1.58) billion in 1980 and have declined since then (Table J1). In 1994, they amounted to ¥ 63 (US$0.62) billion, and for 1996, some ¥ 6 (US$0.59) billion have been budgeted. This steady decline in budgetary expenditures for a diverted area that has remained stable or even increased, is explained by a decrease in per-hectare payment rates, but also by the growing proportion of land converted to orchards, woods and other permanent uses, which until 1996 were included in the diverted area reported for subsequent years, even if payments were no longer received. The decrease in the payment rate has been attributed to budgetary constraints. The administrative expenditures in connection with the PFFRP have been estimated at around 10 per cent of total subsidy expenditure.

Environmental assessment

With respect to supply control, the Japanese land diversion schemes have largely met their declared objectives. Almost all rice producers have participated in the programmes and the actual area diverted has exceeded the national target area in almost each year since 1978. The schemes have sufficiently reduced the supply of rice from domestic production so that no rice exports have been necessary since 1984.

With respect to the environmental effects of paddy field diversion, little information is available. Since the programmes did, until recently, not specify any environmental target, no system for monitoring and measuring their environmental impacts has been implemented. The potential environmental benefits obtained by providing subsidies for certain land use changes can only be assessed in a conjectural way based on the general relationships between the alternative types of land use and their environmental characteristics outlined in Table J2, and several field-level studies about the environmental effects of paddy fields.

Field research into the environmental effects of paddy field production suggests that paddies have the capacity to mitigate the risk of floods by retarding water runoff and reducing peak water discharge from rivers after heavy rainfalls; that they contribute to recharging groundwater resources by temporarily storing rainfall and increasing water infiltration into the soil; and that they can reduce soil erosion by interrupting the flow of surface water on steep slopes (Japanese Ministry of Agriculture, 1995).

A study on the implications of land use changes on paddy fields in an area with high rainfall intensity (3 000 mm per year) and steep slopes found that annual soil loss due to water erosion was lowest on fields that were continuously used for rice production (Iwama and Otsuka, 1995). Shifting from rice or upland crop production to grass-based uses resulted in a fourfold increase in the erosion level. The highest erosion levels were recorded where paddy fields were abandoned. Another study suggests that abandonment of small terraced paddies in hilly areas can significantly increase peak flood discharge after rainfalls and increase the frequency of flooding (Japanese Ministry of Agriculture, 1995). These studies suggest that the diversion of paddy fields from rice production in hilly areas can lead to environmental degradation if appropriate conservation measures are not implemented.

By encouraging alternative uses of paddy fields, the environmental provisions of the diversion programmes are intended to prevent the negative environmental effects of abandoning the fields. To the extent that the alternative uses encouraged by the provisions, including "managed" idling and flooding, have helped preserve the hydrological functions of paddy fields or introduced land uses of comparable conservation value, the risk of soil erosion, landslides and flooding has been reduced. However, in the absence of more precise quantitative information it is difficult to assess how the soil and water conservation value of certain alternative uses, such as forests and orchards, compares to that of paddy-field rice production.

Rice paddy field diversion has also led to shifts in production to other crops, such as flowers and vegetables, and the implications of this shift for the intensity of fertiliser and chemicals use on diverted land would need to be analysed more closely, especially in the light of the new policy objective of reducing environmental pressure on this land. The NPAPP, unlike former programmes, urges farmers to make efforts to reduce environmental pressure on diverted land. The NPAPP is also closely linked to the *Sustainable Agriculture Promotion Plan*, and through this linkage incites farmers to reduce environmental pressure on diverted land as well as on rice-producing land.

Given the increasing importance of the environmental aspects of the programme, it would appear that there is a strong need for a monitoring system and the development of indicators that could be used to evaluate, in quantitative terms, the environmental effects of the schemes. This would also involve a more systematic evaluation of the positive environmental externalities of paddy fields, across farming regions, climatic zones and topographic conditions, and a comparison of the environmental effectiveness of alternative land use options.

As currently applied, the environmental provisions of the paddy field diversion programmes are still characterised by a fairly low degree of environmental targeting. There are no requirements for farm- or field-level assessment of the environmental risks of land diversion, and the conservation measures adopted by farmers have only recently been linked to specific environmental objectives. It is likely that more accurate environmental targeting would increase the effectiveness of the environmental provisions of the programmes, while recognising the limits to integrating, in an effective way, environmental concerns into agricultural policy.

Concluding remarks

The Japanese paddy field diversion programmes are supply control measures that operate by taking land out of rice production. If the diverted land were abandoned, the paddy fields would degrade and their environmental benefits with respect to flood control, soil erosion and landslide prevention would be diminished. The environmental provisions attached to the programmes are aimed at avoiding such degradation by paying farmers to manage diverted paddy fields in environmentally sound ways. This includes appropriate cropping alternatives and/or maintenance of idled paddy fields. Conversion of paddy fields to forests and other perennial crops is also subsidised. With the introduction of soil-improving and landscape-enhancing crops on diverted land, an attempt has been made to strengthen the environmental elements of the programmes.

The environmental provisions are mostly preventive and limited in scope. They did originally not address some of the more important environmental problems in Japanese agriculture, such as water pollution through inappropriate fertiliser and pesticide use. This was reflected by the absence of any reference to agricultural input use on diverted paddies. With the implementation of the NPAPP in 1996, environmental management of diverted land has been formally adopted as a policy objective, and farmers have been encouraged to reduce fertiliser and chemicals use on this land. Requirements to take environmental conditions into consideration in allocating the national land diversion target area across prefectures and municipalities, or in selecting set-aside plots on farms, do not exist. However, municipalities and farm co-operatives can influence the choice of set-aside land in the process of approving the farmers' land diversion plans.

Although the attention paid to the environmental aspects of paddy field diversion has increased in recent years, monitoring and assessment of the actual environmental effects of the programmes are lagging behind. In principle, it can be expected that the environmental provisions have contributed to preventing degradation of diverted land where this land has been used in accordance with the environmental requirements specified by the government. However, there are no quantitative indicators to measure these effects. The development of such indicators is a precondition for an in-depth evaluation of the environmental impacts of paddy field diversion, including the conservation effects of the various alternative paddy field uses.

The environmental provisions constitute, in a pragmatic way, an attempt to integrate environmental concerns into agricultural policy. As a result, a combination of price support, supply control and environmental protection measures exist in the Japanese rice sector. While this reflects the multiple goals of agricultural and environmental policy, it also illustrates the difficulties that arise with respect to achieving a satisfactory degree of environmental targeting. A broader assessment of the effectiveness of the paddy field diversion programmes would require the exploration of alternative approaches to supply management and environmental conservation in the rice sector.

6. SWITZERLAND'S LAND DIVERSION PROGRAMMES

Background

Switzerland's land diversion programmes originated in the early 1990s as part of the government's attempt to halt the trend to increasing production and yields. In 1991, a programme for the "orientation of crop production and extensive farming" was enacted to encourage changes in land use and reductions in the intensity of input use ("extensification"). Land diversion measures under this programme included shifts from cereals production to grassland; temporary land set-aside under a green cover ("green fallow"); preservation of areas of high ecological value, such as hedges, wooded farmland and buffer areas alongside forests, rivers and roads; and the production of renewable resources for non-food consumption.

In 1993, Swiss agricultural policy took a new direction characterised by the gradual dissociation of income and price support. Although Switzerland still has one of the highest levels of market price support among OECD countries, the prices of major agricultural commodities were lowered and direct payments were introduced to partially compensate farmers for the ensuing income losses. At the same time, measures were implemented that laid the basis for payments to farmers in return for the provision of environmental services.

The 1993 agri-environmental measures entitle farmers to receive government payments if they apply land use practices that are deemed beneficial to the environment; shift to "organic" or "integrated" farming systems; and use environmentally-friendly livestock production methods. Among the land use practices eligible for government support are "extensive" pasture, "low-intensive" pasture and floral meadows.

The programmes

The Swiss approach to environmental management in agriculture is character-ised by the integration of land diversion measures and extensification measures. The measures that conform most closely with the definition of land diversion schemes used in this document include payments for green fallow, floral meadows, set-aside of areas of high ecological value, the conversion of arable land to exten-sive pasture, and the production of raw materials for industrial uses. Payments for extensive and low-intensive pasture land could be classified either as land diversion or as extensification measures, as they are granted irrespective of whether the current practice is the consequence of a land use change in the past (land diver-sion) or a continuation of traditional grassland farming with a lower level of input use (extensification). The distinction between land diversion and extensification becomes entirely arbitrary in the case of shifts to organic or integrated farming system, as these usually involve reorganisation of the entire farming operation.

In practice, the situation is even more complex as many farms participate in several programmes at the same time, and the environmental effects of a given measure depend on the other measures in place. For instance, if part of a farm's land base is enrolled in the extensive pasture scheme, the overall effect on chemi-cals use will depend on whether the farmer has the possibility to intensify input use on the remaining land. If the farm is also enrolled in the organic or integrated farming programme, such intensification will not be possible.

The linkages between the different measures have to be taken into consider-ation in an assessment of the Swiss land diversion schemes. The policy measures discussed in this paper include the *ecological compensation measures* of the 1993 policy package (payments for extensive and low-intensive pasture, floral meadows, areas of high ecological value, and the conversion of arable land to extensive grassland), *green fallow*, and the *production of non-food crops*. These measures cover the major policy incentives for land diversion, including those provided by the extensive and low-intensive pasture schemes.

Ecological compensation measures have to be implemented for a minimum duration of six years. Use of chemical plant protection products on enrolled land is forbidden. In addition, fertilisation is prohibited on *extensive pasture land, areas of high ecological value* and *arable land converted to grassland*. A late-season cut of the green cover is prescribed on these lands. On *low-intensive grassland*, nitrogen fertilisation is permitted, but only by way of animal manure. *Floral meadows* have to be seeded with a mixture of wild indigenous plants and must remain free of fertiliser applications. Except for a biyearly cut, no use of the plant cover is permit-ted. Support for floral meadows is restricted to arable land that has been withdrawn from crop production and, like support for the *conversion of arable land to grass-land*, is provided only in the lowland parts of the country. In cases where land of high ecological value (great diversity of the flora and fauna) is involved, individual contracts can be signed between farmers and the cantonal and federal authorities, and supplementary aid is granted.

The **green fallow programme** is targeted to arable land in the lowland areas of the country. Green fallow land has to be withdrawn from crop production for at least one growing season and put under a plant cover. The cover has to be established using a recommended seed mixture, unless natural greening furnishes a cover of adequate plant composition. No fertiliser or chemicals applications are permitted, and no economic use of the grass cover is allowed. Payments are limited to a maximum of 15 hectares per farm and can be received only once every four years for the same plot of land.

Support for the production of **agricultural raw materials for non-food** use is still in its experimental stage. The total supported area is limited to 3 000 hectares. The project's aim is to identify those renewable raw materials that are best suited, from an ecological point of view, for production and industrial use. The major supported crops are rapeseed and miscanthus. Payments are only granted if a contract between the farmer and the user of the raw material exists. Any food or feed use of the raw material or the processed product is forbidden. Non-food crops have to be grown using methods that are "respectful of the environment".

The payment rates of the Swiss land diversion schemes are based on the principle that participation in agri-environmental schemes should not affect farm income. The payments are intended to compensate farmers for the cost increases and/or revenue losses associated with abandoning conventional production on part of their land. In practice, however, there is little differentiation in payment rates by type of farm, agri-environmental measure, or region. A single rate of SF 3 000 (US$2 537) per hectare is applied to four of the six measures: green fallow, floral meadows, the conversion of arable land to grassland, and the production of non-food crops. This rate is the same for all farms and regions where the programmes are offered, and has not changed since 1993 (Table S1).

The payment rates for extensive and low-intensive pasture range from SF 300 (US$254) to SF 1 200 (US$1 015) per hectare depending on the region in which the land is located. The lower rates apply to farmland in the mountains, the higher rates to the more fertile lowland areas. The payment rate differentials reflect differences in the income-generating potential of the land. Land of high ecological value is eligible for an extra premium, which is financed jointly by the Swiss Confederation and the cantons. The magnitude of this premium varies according to the diversity of the flora and fauna preserved on the land.

Participation in the land diversion schemes has increased each year since the introduction of the programmes. In 1994, about 57 000 hectares, or 5.5 per cent of the agricultural land, were enrolled in a land diversion scheme. Of this, almost

Table S1. **Per-hectare payment rates of the Swiss land diversion schemes**

Swiss francs per hectare

Policy measure	1993	1996
Extensive pasture	$800^1/600^2/450^3$	$1\ 200^1/700^2/450^3$
Low-intensive pasture	$600^1/450^2/300^3$	$600^1/450^2/300^3$
Green fallow	3 000	3 000
Arable land converted to extensive grassland	3 000	3 000
Floral meadows	3 000	3 000
Renewable resources for non-food use	–	3 000

1 SF = 0.68 US$ in 1993 and 0.85 US$ in 1995.
1.   All regions other than mountain zones 1-4.
2.   Mountain zones 1 and 2.
3.   Mountain zones 3 and 4.
Source: Swiss government, 1996.

33 000 hectares were accounted for by extensive and low-extensive pasture (Table S2). Around 30 per cent of Swiss farms participated in the extensive pasture scheme and 25 per cent in the low-intensive pasture scheme. By comparison, only 4 per cent of the farmers received payments for the conversion of arable land to grassland, and 2 per cent for green fallow. The lowest participation rates were recorded for floral meadows and the production of non-food crops. Only a few hundred farmers signed up for these programmes, with an average enrolled area of less than one hectare per farm.

Table S2. **Farmer participation in the Swiss land diversion schemes, 1994**

Policy measure	Number of farms	Per cent of total farms	Area (hectares)	Per cent of total agric. area
Extensive pasture [1]	21 032	30.6	22 206	2.1
Low-intensive pasture	16 944	24.7	30 428	2.8
Green fallow	1 504	2.1	2 212	0.3
Extensive pasture on set-aside land	2 580	3.8	2 003	0.2
Floral meadows	176	0.4	77	n.s.
Renewable resources for non-food use	322	0.5	266	0.1

1.   Including land of high ecological value; n.s.: Less than 0.1 per cent.
Source: Swiss government, 1996.

Apart from the land diversion schemes, the programmes for *organic* and *integrated farming systems* have also met with a strong response by farmers. Participation in these programmes increased from 16 per cent in 1993 to over 50 per cent in 1995. This upward trend is expected to continue and by 1998, as much as 90 per cent of the total agricultural area could be farmed in accordance with organic or integrated production methods. This development could contribute to further increases in land diversion, as farmers taking part in the organic and integrated farming programmes have to reserve 5 per cent of their land base for ecological compensation measures.

The land diversion programmes have had a noticeable effect on cereals supply. Government estimates for 1994 suggest that the combination of land diversion and extensification measures in the crop sector reduced cereals production by between 6 to 7 per cent in that year. Around one-third of this reduction was attributed to land diversion, the other two-thirds to extensification measures.

Government expenditures on land diversion programmes amounted to SF 41 (US$35) million in 1994. More than two-thirds of the land diversion payments was accounted for by the extensive and low-intensive pasture schemes, 16 per cent by green fallow and 14 per cent by the conversion of arable land to grassland. Support for floral meadows and the production of non-food crops amounted to around SF 1 (US$0.85) million.

Budgetary expenditures on agri-environmental measures are rapidly increasing and could soon become the single largest group of direct payments to farmers. In 1993, the year that marked the large-scale introduction of agri-environmental measures, some SF 55 (US$47) million were spent on these measures. By comparison, SF 900 (US$761) million are budgeted for 1999.

Environmental assessment

Owing to the recent implementation of the Swiss land diversion programmes, no systematic assessment of their environmental effects has yet been undertaken. The Swiss government has embarked on a comprehensive evaluation project, which focuses on several criteria, including soil nitrogen and phosphorus, use of plant protection products, and biodiversity. By mid-1996, no results of the project were yet available.

In the absence of confirmed empirical estimates, the environmental assessment of the land diversion programmes has to be limited to a discussion of the programme criteria and the results of various research projects. These suggest that the **ecological compensation measures** increase the diversity of the flora and fauna in rural areas and enhance the cultural landscape. In particular, land set-aside in strategic locations, such as buffer zones alongside rivers, lakes, forests and roads, and the conversion of such land into extensive grassland, have beneficial effects for water quality and wildlife.

Enrolment of part of a farm's land base in a set-aside scheme reduces the land available for conventional cropping. This can lead to the more intensive cultivation of the remaining land, unless the farm participates at the same time in the organic or integrated farming programmes, which subject the entire farm to strict environmental management.[18] However, with the majority of farms expected to participate in either the organic or the integrated farming programme in the near future, the risk of input substitution will diminish.

The **production of raw materials for non-food use** is currently subject to an in-depth evaluation, which is based on a comparison with equivalent products (for example, the same amount of energy produced from miscanthus or from fuel-oil), and which takes into account all resource and energy flows necessary for extraction, production, transportation, utilisation and waste disposal. According to preliminary results, the environmental impact associated with renewable materials is often smaller than that of equivalent products (Office fédéral de l'agriculture, 1995). However, eutrophication of water bodies is systematically more severe due to greater nitrate and ammonium emissions. Moreover, combustion of agricultural raw materials for energy production can increase acidification and create ground-level ozone. Use of renewable materials for manufacturing purposes is in this respect preferable to use for energy generation. The environmental impact of producing renewable raw materials does thus not only depend on the farm-level effects, but also on the final products.

Concluding remarks

The Swiss land diversion programmes pursue the double objective of supply control and environmental conservation. While the early programmes, implemented in 1991, were primarily aimed at shifting land from cereals production to alternative uses, the measures introduced in 1993 were developed within a broader framework for agri-environmental management in agriculture. Within this framework, a "menu" of measures encouraging changes in land use, reductions in fertiliser and chemicals applications, and the adoption of integrated and organic farming methods have been implemented. Land diversion measures are an integral part of this package.

The environmental conditions for set-aside land are fairly restrictive, and in most cases involve an outright ban on the use of commercial fertilisers, chemicals for weed control, and plant protection products. They also require establishment and maintenance of an effective green cover, either through natural regeneration or by seeding the land to a recommended seed mixture. These seed mixtures, which are composed of local plant species, are intended to preserve or revive the local flora. In the case of establishing floral meadows on former cropland, an effort has been made to reintegrate into mainstream agriculture a broad array of plants that were once part of the cultural landscape.

There are indications that the land diversion schemes have benefited the environment and landscape, yet the magnitude of these benefits are not known. The Swiss government is engaged in a comprehensive evaluation project, but no results were yet available by mid-1996. Major benefits are expected in the area of biodiversity, landscape, nutrient run-off and farm chemicals use. The environmental effects are likely to differ by policy measure. Floral meadows, for instance, can be expected to enhance the landscape, increase plant diversity, provide feed and shelter for wildlife, and reduce fertiliser run-off. The production and utilisation of non-food crops, on the other hand, could have a more mixed record. It might have helped conserve non-renewable resources in energy production and manufacturing, but it could also have increased acidification and water pollution.

Farmers who participate in the schemes receive government payments to compensate for the income losses associated with land diversion. In most cases, a uniform payment rate per hectare is applied, although differences by land use type and region exist. Regarding ecological compensation measures, supplementary aid can be granted in addition to the base payment where land uses are involved that are particularly beneficial for plant and animal life. Moreover, in well-head areas and sensitive buffer zones, set-aside with green cover is eligible for an extra payment. In addition, many local communities provide incentives for site-specific targeting of ecological compensation measures. These supplementary payments encourage enrolment of environmentally sensitive land in the programmes.

Expenditures on land diversion programmes and other agri-environmental measures are projected to increase as Swiss agricultural policy continues to shift from price support to taxpayer-funded direct payments. To justify these expenditures and ensure that the funds are directed to their most effective use, more information on the environmental effects of the programmes will be required. The development and implementation of agri-environmental indicators could be a useful step in this direction.

In the future, land diversion is likely to take place to a larger degree on farms participating in the organic and integrated farming programmes, as these farms are required to set aside a certain proportion of their land for ecological compensation measures. If the current trend continues, the large majority of Swiss farms will participate in one of these programmes by the end of the decade. The integration of a land diversion requirement into organic and integrated farming is an example of the complementarity of Swiss agri-environmental measures.

The Swiss land diversion programmes have been implemented in a context of high market price support. By encouraging intensive production, the high levels of price support may have increased cereals output and created pressure on the environment. The current land diversion programmes may to some extent merely undo some of the production and environmental effects of price support. Progressive agricultural policy reform, centred on a lowering of market price support, could reduce the harmful effects of support on the environment and facilitate a better targeting of land diversion measures to situations where the market fails to provide the environmental benefits demanded by society.

7. THE UNITED STATES' CONSERVATION RESERVE AND ACREAGE REDUCTION PROGRAMS

Background

The United States has a long history of using land diversion schemes as instruments of supply control and environmental conservation. The *Conservation Adjustment* and *Agricultural Conservation Programs*, which were in place from 1933 to 1948, idled more than 40 million acres (1 acre = 0.4 hectares) of cropland in their peak years. Between 1956 and 1971, substantial areas of land (up to 30 million acres) were diverted from production under the *Soil Bank* programme. Much of the land idled in the Soil Bank was eventually returned to cropping (Office of Technology Assessment, 1995). The two most important land diversion programmes in recent years are the Conservation Reserve Program and the Acreage Reduction Program. Annual *Acreage Reduction Programs* were operated from 1961 to 1996, and the *Conservation Reserve Program* has diverted land from agricultural production since 1986.

Programme description

The **Conservation Reserve Program** (CRP) was established with the Food Security Act of 1985 as a voluntary long-term cropland retirement programme. Farmers who participate in the CRP commit themselves to retire highly erodible or environmentally sensitive cropland from production for 10 to 15 years, and to keep it under a permanent cover, such as grass or trees. In exchange, they receive annual rental payments and half the cost of establishing the vegetative land cover.

The primary stated goal of the CRP is to reduce soil erosion on highly erodible cropland, yet other environmental and non-environmental objectives are also pursued. From the inception of the programme, the secondary objectives have included: preserving soil productivity, improving water quality, reducing off-site soil sedimentation, creating wildlife habitat, curbing surplus production, and providing income support to farmers. While the list of environmental objectives has largely remained the same, the relative importance accorded to each of these objectives has changed as the programme has evolved.

Nine CRP sign-up periods were organised under the 1985 Food Security Act between March 1986 and July 1989. The criteria for land eligibility were based on three indicators: a high level of *actual* erosion relative to the *soil loss tolerance level*;[19] a high level of *potential* erosion as measured by the *erodibility index*;[20] and the suitability of the land for cropping as measured by *land capability class*. A parcel of land was declared eligible if at least two-thirds of its area met these criteria.

Roughly 101 million acres, or almost one-quarter of total US cropland, met the definition of highly erodible land (Osborn *et al.*, 1992). In addition, cropland adjacent to water bodies or subject to scour erosion, and certain cropped wetlands were included in the pool.[21] However, to prevent a regional concentration of CRP land and mitigate the effects of the programme on agro-food industries and rural employment, enrolment was limited to 25 per cent of the cropland of a county, a restriction that reduced the area of eligible land to around 70 million acres. The enrolment goal to be attained by 1990 was set at 40-45 million acres.

Implementation of the CRP was based on an "offer system". The government established a single maximum rental rate for all land in a given area and farmers could, at this rate, enrol as much of their eligible land as they wanted. Areas with uniform payment rates comprised one or several counties, or even an entire state. Nearly 34 million acres of cropland were enrolled in the CRP during the first nine sign-up periods.

The 1990 Food, Agriculture, Conservation, and Trade Act extended CRP enrolment through 1995. At the same time, the objectives of the programme were revised, with greater emphasis being placed on improving water quality and wildlife habitat. Planting of hardwood trees and conversion of existing CRP land to trees was also encouraged. The Chesapeake Bay, the Long Island Sound and the Great Lakes regions were designated as conservation priority areas. The enrolment goal of 40-45 million acres was initially retained but later revised downwards to 38 million acres. This figure included land enrolled in both the CRP and the newly created Wetland Reserve Program.

The bidding process was also modified. For each eligible parcel of land offered by farmers, the expected environmental benefits and estimated CRP expenditures were evaluated, and the benefit/cost ratio was used to rank the bids. The environmental benefits index included: surface and groundwater improvement, preservation of soil productivity, assistance to farmers most impacted by conservation compliance, encouragement of tree planting, and enrolment in conservation priority areas and areas identified under the Water Quality Initiative (United States Department of Agriculture [USDA], 1994*a*). The cost component was based on the annual government payment requested by the farmer and the estimated cost-share expenditure to establish permanent cover.

The uniform payment rate ceiling for a given area was replaced by individual payment limits for each tract of cropland, which were adjusted for soil productivity to reflect the estimated opportunity cost of diverting the land from crop production. Bids were only accepted if the annual payment requested by the farmer did not exceed the specified limit (Osborn, 1995). In order to maximise competition among farmers, neither the payment limits nor the exact composition of the environmental benefits indicator was disclosed by the government.

Three enrolment periods were organised under the revised rules between 1990 and 1992. For the last of these sign-ups, additional conservation priority areas were designated based on proposals made by several states (USDA, 1992). Bids involving specifically targeted land, such as well-head protection areas or land on which windbreaks or filter strips would be installed, were automatically accepted. For all other bids, those with the highest benefit/cost ratios were accepted until the allocated funds were exhausted. Two and a half million acres of land were enrolled during the three sign-up periods.

After the 12th sign-up, in June 1992, new enrolment was halted due to Federal budget pressure, although the enrolment goal had not been reached. A total of 375 000 contracts had been signed and 36.4 million acres, or around eight per cent of US cropland, had been converted to conservation uses under the CRP.

The first contracts, covering approximately two million acres of land, were scheduled to expire at the end of September 1995, and an additional 22.4 million acres in 1996 and 1997. With the future of the CRP depending on the 1996 Farm Act, and therefore still uncertain in early 1995, an interim solution was offered by the government. Farmers with expiring CRP contracts were given the option to extend their contracts by one year. Extensions were requested by farmers on about 78 per cent of the eligible acreage.

In addition to the one-year extension, CRP participants were offered the opportunity to terminate their contracts prematurely or reduce the enrolled area without penalty.[22] However, farmers who opted for an early release and returned CRP land to crop or pasture production were obliged to manage this land according to a Basic Conservation System plan[23] or an approved haying or grazing plan. Furthermore, for the first year after early release they were not entitled to deficiency payments on former CRP land. Partly because of these restrictions, fewer farmers than expected took advantage of the opportunity to terminate their contracts and only 651 000 acres were returned to production (Osborn, 1995).

Concurrently with the early release, a 13th CRP sign-up period was organised in September 1995 to replace the land thus exiting the programme. The eligibility criteria for cropland were tightened and the environmental benefits index used for ranking the bids was modified. A parcel of land could be offered for enrolment if it had an *average* erodibility index of greater than eight. This replaced both the land capability classification and the requirement of high erodibility on two-thirds of the parcel. In addition, land at risk of flooding, well-head protection areas, riparian buffer zones, small farmed wetlands, and cropland located in conservation priority areas were admitted in the bidding process irrespective of the level of erodibility.

The index used in ranking the bids consisted of five criteria, of which four represented the expected environmental benefits of the land parcel and the fifth the budgetary cost of enrolment. The four environmental components were: water quality protection, habitat creation, soil erosion abatement, and tree planting. Special areas, such as filter strips, shallow water areas, windbreaks and shelter-belts automatically received the maximum environmental scores to favour their enrolment. In addition, a 10 per cent rental bonus was offered for filter strips along waterways (Osborn, 1995).

Contrary to the procedure used in previous sign-ups, the government informed each applicant of the maximum bid price that would be accepted for the land offered. Farmers could bid at this rate or at any rate below, knowing that a lower requested payment rate would increase the probability of acceptance. The additional area enrolled was limited to the number of acres returned to production under the early release option. It was believed that the more selective land acceptance criteria would lead to the replacement of exiting land by more environmentally sensitive land and increase the degree of environmental targeting of the programme.

The Federal Agriculture Improvement and Reform Act of 1996 extended the CRP through the year 2002. A maximum of 36.4 million acres can be enrolled at any one time. CRP contracts that were signed before 1995 and have been in effect for at least five years can be terminated before their date of expiration, provided that they do not involve lands rated high in environmental value. Based on these criteria, an estimated 7 to 8 million acres of CRP land were eligible for early release in spring 1996, although it was expected that farmers would remove only around one million acres from the programme under this option. In addition to early release, CRP contracts covering some 16.9 million acres of land are scheduled to expire on a regular basis by 1 September 1996.

CRP land that has a crop acreage base history can be entered into "production flexibility contracts" under the 1996 Farm Act when the CRP contracts expire or are terminated. An estimated 50 per cent of 1996 CRP land has an established farm programme base. Expired or terminated contracts may be replaced by new enrolments, funded with money saved on withdrawals. More stringent environmental criteria than in the past will apply to new land sign-ups.

The **Acreage Reduction Program** (ARP) was instituted in order to limit the production effects of agricultural price support and was the major supply control measure in crop production until its elimination in 1996. Under the ARP, farmers participating in commodity programmes were required to set aside a specified percentage of their area base in order to receive programme benefits. Crops under the ARP included wheat, barley, corn, sorghum, oats, cotton and rice. A farmer's area base for wheat and feed grains was the average of the area planted or "considered planted" to the crop during the previous five years; the acreage base for rice and cotton was based on averages of the last three years. Areas that were "considered planted" included those that had been diverted under past commodity programmes and "flex" acres planted to other crops (USDA, 1995b).

Acreage reduction percentages were announced in advance of each crop year. They were commodity-specific and were determined as a function of commodity stocks and, since 1990, domestic demand and export levels.[24] Consequently, the accumulation of stocks in the second half of the 1980s resulted in high set-aside requirements (attaining levels in excess of 25 per cent for major crops) and the declining stocks of the early 1990s led to low set-aside requirements. In the case of wheat, for instance, the set-aside percentage was reduced to zero in the 1993/94 crop year and remained at this level until the programme was eliminated.

The set-aside land had to be placed in soil conserving use for the growing season. However, no specific requirements regarding establishment of a land cover, choice of cover plants, or other land management practices existed. Moreover, there were no incentives to divert the most erodible or otherwise environmentally sensitive land, or to keep the same plots of land out of production for a prolonged period.

Environmental assessment of CRP and ARP

Assessment of the land diversion criteria

The primary environmental objective of the **Conservation Reserve Program,** soil erosion control, competes with other objectives, such as reduction of water pollution and improvement of wildlife habitat. In the earlier sign-up periods, the criteria for land eligibility were designed to target areas with high levels of actual or potential soil erosion. In later sign-ups, attempts were made to evaluate the multiple benefits of a given parcel of land and accept the bids with the highest benefit-cost ratios.

In particular, the introduction in 1990 of an environmental benefits index has reduced the focus on erosion control and shifted attention to the negative off-site effects of crop production. This tendency has been reinforced by the increasing importance given to conservation priority areas. The move to payment ceilings based on soil productivity has created an opportunity to bring government payments more in line with the actual income foregone by not producing grains on diverted land. Finally, the ranking of bids based on the ratio of expected benefits to expected budgetary outlays is likely to have moved the programme in the direction of greater environmental effectiveness and cost-efficiency.

Among the participation criteria that may have restricted efficiency gains is the 25-per cent cropland enrolment limit at the county level, which has effectively eliminated large areas of environmentally highly sensitive land from being considered for CRP enrolment (United States General Accounting Office [GAO], 1989). The policy of "no economic use on CRP land" may have kept annual payment rates unduly high and increased the dead-weight loss of the programme. It may also have prevented contract holders from developing alternative, economically sustainable uses on CRP land. In the case of tree plantings, which have been encouraged since the beginning of the programme,[25] economic benefits are eventually obtained, yet no difference was made in the payment rate compared with land put under a grass cover.

Under the **Acreage Reduction Program,** participants in grain programmes were required to set aside a proportion of their land, irrespective of whether or not this land was environmentally sensitive. It is likely that such land was only diverted if it happened to be economically least profitable, and then only if farmers chose to participate in government support programmes, and the set-aside requirements were high enough to cover the land in need of protection.

Farmers could keep the same parcels out of production in successive years or rotate set-aside plots. Where land was retired for only short periods, any environmental benefits were limited, even if appropriate conservation measures were taken. Where the same plots were retired for a number of years, opportunities for effective conservation could have been created. However, investments in conservation measures would have been effective only in periods with stable set-aside percentages, which were rare.

In principle, the ARP required that the set-aside land be put to conserving uses. However, no cover crop requirement existed and the permitted conserving uses included summer fallow, which has been demonstrated to be harmful to the environment (Gray and Paddock, 1993). Moreover, compliance with the conservation use requirement was difficult to monitor.

Effects on land use, input use and farming practices

From 1984 to 1995, more than 55 million acres of cropland were on average idled under annual land diversion programmes and the CRP (Table US1). In the later years, the CRP accounted for more than one-half of the total area diverted from production.

By 1995, the **Conservation Reserve Program** had diverted 36.4 million acres of highly erodible or environmentally sensitive land from crop production. This corresponded to eight per cent of US cropland. Approximately two-fifths of CRP land are located in the Northern and Southern Plains, and one-third in the Mountain region and the Corn Belt (Table US2). The regions with the largest areas enrolled in the 1986-89 period were the Northern and Southern Plains and the Mountain region. After 1990, the largest enrolments were achieved in the Corn Belt and Lake

Table US1. **Cropland idley by Federal farm programmes**

Million acres

Year	Annual programmes[1]	CRP	Year	Annual programmes[1]	CRP
1984	27	0	1990	28	34
1985	31	0	1991	30	35
1986	46	2	1992	20	35
1987	60	16	1993	23	36
1988	53	25	1994	13	36
1989	31	30	1995[2]	14	35

1. Including base acres idled under the "0/92" and "50/92" provisions from 1986 through 1992, but excluding base acres signed up under these provisions and planted to oilseeds in 1991, 1992, and 1993.
2. Preliminary estimate.
Source: USDA, 1995e.

Table US2. **Regional distribution of CRP land**

Région	1986-89		1991-92	
	Land enrolment (thousand acres)	Erosion reduction (tonnes/acre/year)	Land enrolment (thousand acres)	Erosion reduction (tonnes/acre/year)
Northeast	200	13	20	6
Appalachian	1 060	26	100	19
Southeast	1 570	15	120	12
Delta States	1 090	19	160	11
Corn Belt	4 730	18	880	15
Lake States	2 630	16	380	10
Northern Plains	9 430	15	230	17
Southern Plains	5 080	32	270	28
Mountain	6 440	19	250	16
Pacific	1 700	13	90	12
United States	**33 920**	**19**	**2 500**	**15**

Source: USDA, 1994b.

State regions. This geographic shift from the Great Plains to the Midwest resulted from the new bid process, which placed more emphasis on water quality improvement in targeting the land, and on the benefit/cost ratio in ranking the bids.

The successive modifications in the bid process have led to a greater geographical dispersion of enrolments in recent sign-ups. More cropland in the eastern half of the United States with higher water quality benefits has been selected, despite higher average bid prices (Benbrook, 1995). Moreover, an increasing proportion of the land accepted into the programme came from specifically targeted areas. In the

12th sign-up in 1991, for instance, around 185 000 acres, or almost one-fifth of the land enrolled, came from conservation priority areas (USDA, 1992). Relatively little additional low-cost erodible land in the western United States was contracted after 1990.

Of the total cropland enrolled, 17.8 million acres were classified as highly erodible in 1982 (USDA, 1995d). Since some of the most highly erodible land was brought first into the CRP, estimated erosion reduction rates on land enrolled in successive sign-up periods declined steadily, from an average of 28 tonnes per acre and year in 1986 to 19 tonnes per acre and year for land enrolled in 1991. By region, the most erodible land was enrolled in the Southern Plains and the Appalachian region, and the least erodible in the Pacific area and the Northeast. The difference in soil erodibility between the earlier and more recent sign-ups is particularly notable in the Delta States, the Northeast and the Appalachian region (Table US2).

Among the factors determining the environmental benefits of the CRP are the previous and current uses of land enrolled in the programme. With respect to previous land use, an indication of the crops grown can be obtained from enrolment statistics concerning land that had been part of a programme crop base before it was placed in the CRP. Over 23 million acres or about two-thirds of the CRP land are from programme bases. Wheat accounts for almost half of this area, followed by corn with more than 4 million acres and barley with almost 3 million acres (Table US3). The remaining 5 million acres came from acreage bases of oats, sorghum and upland cotton.

Table US3. **Areas enrolled in the CRP from programme crop bases**

Million acres

Wheat	Corn	Barley	Oats	Sorghum	Cotton	**Total**
10.8	4.3	2.8	1.4	2.5	1.4	**23.2**

Source: USDA, 1995e.

Conservation methods under the CRP reduce erosion rates primarily by maintaining a protective vegetative cover over the soil. Roughly 87 per cent of the CRP land has been seeded to grass. But the Conservation Reserve also contains 2.5 million acres of trees, 2 million acres of land managed in line with special wildlife practices, and 410 000 acres of wetlands (USDA, 1994a).

The area planted with trees is of interest not only because soil erosion rates in forests are very low (Pimentel *et al.*, 1995), but also because woodlands are less likely than grassland to be returned to production after the CRP contracts expire. Because of the extended time frame of forest plantings and the favourable conditions for wildlife and ecosystem development offered by woodlands, this type of alternative land use can have potentially large environmental benefits.

More than half the afforested CRP land is in the Southeast, where second-growth forests for commercial biomass production have increased considerably since the 1950s (Dunn *et al.*, 1993). Thirty per cent of land conversion to trees took place in the Delta States. Very little tree planting occurred in the Great Plains, the Corn Belt and the Pacific regions. Overall, six per cent of the cropland enrolled in the CRP before 1990 and twelve per cent of the land signed up after 1990 were turned into woodlands.

Since 1988, farmers have had the option of enrolling land within 66 to 99 feet of a waterway regardless of the degree of erodibility. Some 5 200 miles of filter strips covering around 42 000 acres of land have been enrolled in the CRP based on this provision (USDA, 1994b). Although this is quite a sizeable area, many farmers are thought to have been reluctant to offer riparian filter strips for enrolment because these lands are often highly productive (Benbrook, 1995).

In several states, farmers have resorted to a large-scale adoption of land management practices that favour the development of wildlife. Such practices include establishment of permanent wildlife habitat, protection of shallow water bodies for wildlife and water fowl, and provision of wildlife feed plants. As much as 27 per cent of CRP land in North Dakota, 19 per cent in Nebraska and 16 per cent in Wyoming are subject to wildlife-improving practices.

Although land set aside under the **Acreage Reduction Program** was supposed to be put into conserving uses, pesticides and herbicides seem to have been routinely applied to the idled land. Since the set-aside requirement could change from year to year, there was no incentive for farmers to invest in multi-year conservation plans. The land use and input use changes caused by the ARP were dependent on the duration of the set-asides, the degree of erodibility of the idled land, and the management practices applied to this land. The lack of systematic information on these variables precludes a comprehensive analysis of the environmental effects of the ARP.

Both the CRP and the ARP have reduced the land base available for crop production. In principle, this might have led to the adoption of more intensive farming practices on the land remaining in annual cropping, including an increase in applications of fertilisers and farm chemicals. Land diversion may even have increased *total* intermediate farm input use if more intensive production on the remaining land has outweighed input savings on the idled land. If this has actually occurred depends to a considerable degree on the ease with which fertilisers and chemicals can be substituted for land in the production process. In cases where applications of intermediate inputs can be easily increased, set-aside could have resulted in a higher level of intermediate input consumption (Hertel and Tsigas, 1991; Tobey and Reinert, 1991).

On the other hand, farmers with a fixed labour base and no off-farm employment opportunities might resort to more labour-intensive tillage practices and less pesticide use in response to land diversions with, as a consequence, less chemicals pollution but perhaps more soil erosion. Fertilisers could be applied more frequently and in smaller portions, thus reducing the total amount of fertilisers used and nutrient leaching. Changes in cropping patterns in favour of more input-intensive crops and regional shifts in production could also influence total input use, although model-based estimates suggest that such shifts might not have been very important (Ribaudo et al., 1994).[26]

The impact of set-aside on total input use depends also on several other factors, including consumers' response to the food price increases caused by land diversion,[27] and the possibility of bringing new land into production. To the extent that higher commodity prices draw pasture land and other land into crop production, the net removal of land from cropping will be smaller than the area enrolled in the land diversion programmes. By 1987, for instance, 15.8 million acres of cropland had been diverted from production under the CRP, but the net removal was estimated to have been only 12.6 million acres (Ribaudo et al., 1990).

Overall, there is so far little evidence of an increase in fertiliser and chemicals use in production in response to land diversion programmes. However, the magnitude of the input substitution effect is an important empirical question which would need to be further investigated.

Effects on the environment

According to USDA estimates, the **Conservation Reserve Program** may have reduced soil erosion on the diverted land by nearly 700 million tonnes per year, or 19 tonnes per acre on average (Table US2). Compared with conditions prior to implementation of the CRP, this would imply a 22 per cent reduction of erosion on US cropland (USDA, 1994b). However, estimates derived from the National Resource Inventory Data indicate that actual reductions in soil loss may have been considerably smaller – in the area of 370 million tonnes annually (GAO, 1995a).

Erosion reduction is associated with direct on-site benefits in terms of improved soil quality and productivity. Soil erosion adversely affects soil productivity by reducing water infiltration rates, water-holding capacity, nutrient levels, organic matter, soil organisms and soil depth. When erosion occurs, the amount of water run-off increases and less water becomes available for the crop. Eroded soil typically contains several times more nutrients and organic matter than the soil left behind. The depletion of nutrient reserves and organic matter undermines the formation of soil structure and the development of soil organisms. One of the most long-lasting effects of erosion is the decline of the topsoil layer. The process of soil formation is very slow and even if erosion were completely stopped it would take hundreds of years to replace the lost soil (Pimentel et al., 1995).

The off-farm environmental benefits of erosion control include reduced damage from siltation and eutrophication of waterways, and less disruption of ecological systems and wildlife habitat. In some cases, erosion may cause greater damage off the farm than to soil productivity and crop yields (Ribaudo et al., 1990). The most serious off-site damages of soil erosion are thought to be caused by soil particles entering the water systems. It is estimated that around 60 per cent of the eroded soil is deposited in streams and rivers (Pimentel et al., 1995). The sediments along with nutrients, fertilisers and residues of farm chemicals pollute water resources for downstream users, harm aquatic plants and reduce habitat quality.

The CRP has increased the area of natural vegetation, with benefits for wildlife, landscape and regional biodiversity. The main beneficiaries are grassland and wet-land habitat, and woodland species. Tree planting on CRP land is likely to have increased the number and size of woodlots, thereby providing the conditions for the development of valuable wildlife habitat both at the forest edge and in the interior.[28] Moreover, tree corridors established between formerly isolated woodlots reduce the dispersal barriers to wildlife and decrease the degree of landscape fragmentation. The afforested areas constitutes an additional carbon sink that can improve the regional carbon balance, especially in the south-eastern United States, where the majority of diverted land is under trees (Dunn et al., 1993).

Other benefits of the CRP include reductions in air pollution from soil particles, and related health risks and damages to buildings and structures, especially in arid regions. Effects of the CRP on groundwater quality through reduced leaching of nutrients and pesticides could also have been important in some areas, although land vulnerable to groundwater pollution was not targeted in the major sign-up periods.[29] In addition, the effects of land diversions on groundwater supplies are difficult to measure.

With respect to the environmental impacts of the **Acreage Reduction Program,** only few observations can be made. The practice of summer fallow on land set aside under the ARP is likely to have prevented significant improvements on that land. Fallowing land after periods of intensive cropping could have disturbed the nutrient balance and trigger a release of surplus nutrients that were prevented from leaching under intensive cultivation. This effect may have been mitigated in situations where the idled land was kept under a vegetative cover.

Budgetary expenditures

Annual rental payments to farmers account for the largest part of budgetary expenditures under the **Conservation Reserve Program.** They have exceeded US$1.5 billion during every year since 1991 (Table US4). Cost-share payments for the establishment of the vegetative cover were only important in the early years of the programme and reached US$285 million in the peak year. Budgetary expenditures on the CRP have accounted for roughly half of total federal government expenditure on agricultural conservation programmes in recent years. Compared with commodity programme outlays, however, they have been relatively modest, amounting on average to less than ten per cent.

Table US4. **Conservation Reserve Program expenditures**

Million dollars

	1986	1987	1988	1989	1990	1991	1992	1993	1994[1]	1995[1]
Rental payments	0	410	760	1 162	1 394	1 590	1 613	1 510	1 729	1 739
Cost-share payments	12	246	285	182	118	41	39	32	14	4
Total	**12**	**656**	**1 045**	**1 344**	**1 512**	**1 631**	**1 652**	**1 542**	**1 743**	**1 743**

1. Enacted.
Source: Office of Technology Assessment, 1995.

Until 1995, federal budgetary expenditures on the CRP were to some extent offset by deficiency payment savings on idled crop base acreage. Adjusted for such savings, the annual net budgetary cost was estimated in 1994 to amount to roughly US$1 billion (USDA, 1994a). As from 1996, there will be offsetting savings from reductions in "production flexibility contract payments".

Average payments per acre ranged from US$42 for contracts entered in the first enrolment period to US$63 in the twelfth sign-up period, with an average of US$50 over all contracts signed in the first twelve sign-ups (Table US5). There have been discussions as to whether the rental payments made for contracts signed before 1990 were on average higher than the actual opportunity cost of the land enrolled, and budgetary expenditures therefore larger than need have been. The offer price system used for the first nine sign-ups was criticised as inflating per-acre payments (GAO, 1989). In several areas, especially in the Southwest, payment rates exceeded the local rental price of farmland. In such cases, the same land could probably have been attracted into the CRP at a lower budgetary cost.

Table US5. **Average *annual* rental payments and estimated erosion reductions under the CRP**

Sign-up period	Payments [1]	Soil saved [2]	Sign-up period	Payments [1]	Soil saved [2]
March 1986	42	26	February 1989	51	14
May 1986	44	27	July/August 1989	51	14
August 1986	47	25	March 1991	54	17
February 1987	51	19	July 1991	60	15
July 1987	48	17	June 1992	63	16
February 1988	48	18			
July 1988	50	17	**Total**	**50**	**19**

1. Dollars per acre.
2. Tonnes per acre.
Source: USDA, 1994*b*.

With the introduction of a competitive bidding process after 1990 and the assessment of bids against productivity-adjusted payment rate limits, the likelihood of paying more than the rental price of similar land in the area decreased. It is estimated that if this process had been applied before 1990, total annual government payments could have been substantially reduced. The gains could have been in the order of 25 per cent (USDA, 1994*a*), or even as high as 38 per cent (Smith, 1995).

The increased cost-effectiveness of the competitive bidding process is not reflected in the average annual payment rates of contracts signed since 1990 (Table US5). In fact, payment rates have increased rather than decreased. Moreover, estimated average annual erosion reduction rates for these contracts are lower than those achieved in the earlier sign-ups. This apparent contradiction is explained by the fact that the pre-1990 land eligibility and acceptance criteria were largely targeted at highly erodible lands and that land with the lowest opportunity cost was offered first for enrolment. The environmental characteristics of the land targeted in the later sign-ups were more diverse and involved land with higher soil productivity and therefore higher rental prices.

Relationship with other agricultural conservation programmes

The United States has implemented a number of government programmes that influence agricultural land use, with potential benefits for the environment. Some of these, such as the "underplanting provisions" of the 1985 Farm Security Act or the *Wetland Reserve Program* created in 1990 have, like the CRP and ARP, retired cropland from agricultural production. Others, such as the *Water Bank* programme and the "sodbuster" and "swampbuster" provisions of the 1985 Food Security Act, protect non-agricultural land from conversion to cropland.

The programmes that call for the temporary or permanent retirement of land from agricultural production, or prevent the conversion of non-agricultural land to agricultural uses, can be distinguished from programmes that are aimed at changing farming practices and technologies used in cropping, without necessarily affecting land use patterns. The latter, which include the *Conservation Compliance Provision* of the 1985 Food Security Act, are not directly within the scope of this study. However, because they are often closely linked to land diversion schemes, either by offering an alternative approach to environmental conservation or by complementing land diversion programmes, they merit brief discussion.

The *"underplanting provision"* of the 1985 Food Security Act, allowed producers of wheat and feed grains who participated in commodity programmes to reduce their plantings in favour of conservation uses, and still receive deficiency payments for a large proportion of their eligible base area. Initially, producers of upland cotton and rice who planted between 50 and 92 per cent of their permitted area to the programme crop and devoted the remaining area to conservation uses, could receive deficiency payments on 92 per cent of their maximum payment acreage (the "50/92" provision).[30] In 1988, the provision was extended to include wheat and feed grain producers, who were allowed to put the entire permitted area into conservation use while remaining eligible for deficiency payments on 92 per cent of the payment acreage (the "0/92 provision). After 1993, deficiency payments could only be received on 85 per cent of the permitted area (the "0/85" and "50/85" provisions). The underplanting provisions were eliminated in 1996. Farmers were permitted to enter the idled land into production flexibility contracts and receive market transition payments, provided that the land is planted to grass or another soil-conserving crop to prevent erosion.

The agricultural **Wetland Reserve Program** (WRP) was established by the 1990 Food, Agriculture, Conservation, and Trade Act. This programme authorises the USDA to pay farmers for the restoration of wetlands currently used for agricultural production. The conversion target, to be reached by the year 2000, was set at 975 000 acres. Eligible lands include restorable wetlands, riparian corridors and adjacent cropland that can serve as buffer zones. A total of 55.6 million acres of cropland converted from former wetlands is eligible for the WRP. Estimates suggest that the largest concentrations of low-cost restorable wetlands are in the Northern Plains and the Corn Belt (Carey *et al.*, 1990).

The restoration of wetlands concerns mostly land that had been converted to agricultural uses under previous federal programmes. In the period from 1954 to 1974, the United States was losing wetlands at a rate of 690 000 acres per year, of which 87 per cent were accounted for by conversion to farmland. In 1990, the federal government committed itself to preventing future net losses of wetlands, and the WRP was introduced to help achieve this goal.

The WRP permits economic uses of the restored wetlands, such as hunting, fishing, haying or grazing, provided that they are compatible with wetland preservation. Participating farmers receive a payment, which must not exceed the market value of the land, and up to 100 per cent of the cost of wetland restoration. About 50 000 acres were accepted into the WRP when the programme was opened for enrolment in 1992. In 1994, a further 75 000 acres were enrolled, and an estimated 118 000 acres were signed up in 1995. The Federal Agriculture Improvement and Reform Act of 1996 extended the WRP through the year 2002. Beginning in fiscal year 1997, the area will be split into three equal proportions: one-third of WRP enrolments will be covered by permanent contracts, one-third by 30-year contracts, and one-third by cost-share agreements for wetland restoration, with cost-share payments ranging from 50 to 100 per cent.

The USDA's **Water Bank** programme, which was authorised in 1970, offers annual per-acre payments to landowners who agree not to destroy enrolled wetland areas for a period of 10 years. Cost-sharing payments are available for installation of conservation practices designed to maintain vegetative cover, control erosion, improve habitat, conserve surface water, and manage bottomland hardwoods. Agreements covering almost 500 000 acres of land at an average rate of US$15 per acre were signed under this programme. Payments to farmers under the Water Bank programme ranged from US$7 million to US$17 million annually between 1983 to 1994. No funds were authorised for 1995 (Office of Technology Assessment, 1995).

The **Wetland Conservation** or **"swampbuster" provision** of the 1985 Food Security Act makes farmers ineligible for farm programme benefits if they drain wetlands for use in farm production. Benefits are suspended until the converted wetland is restored. The swampbuster provision does not entirely stop wetland conversion to cropland, as it relies on farmer participation in farm support programmes. By 1992, nearly 700 farmers had lost programme benefits because of violation of the swampbuster provision, involving about 7 500 acres of wetlands. The swampbuster provision may have had a significant part in the reduction of agricultural wetland conversion from 240 000 acres annually from the mid-1970s to the mid-1980s to 15 000 acres per year in the 1987-91 period (Heimlich and Melanson, 1995). Wetland losses due to non-agricultural activities have in recent years outpaced those caused by agriculture (USDA, 1994b). The **"sodbuster" provision** requires that farmers who convert highly erodible land to commodity production implement an approved conservation system in order to be eligible for programme benefits (USDA, 1995b).

The **Conservation Compliance Provision** (CCP) of the 1985 Food Security Act promotes environmentally sound farming practices on highly erodible land in intensive crop production. Farmers are required to comply with an approved conservation plan, which had to be implemented by the end of 1994, to remain eligible for farm programme benefits. The conservation plan has to achieve "substantial" reductions in soil erosion. Investments necessary to implement the conservation measures can be co-funded by the government. An estimated 143 million acres of cropland with an erodibility index of greater than eight are subject to the compliance requirement.

More than one million conservation compliance plans were developed by producers covering approximately 100 million acres of highly erodible land. Ninety-five per cent of those plans were implemented by the legislated deadline. Since the conservation plans have to achieve "substantial" reductions in soil erosion but not necessarily sustainable erosion levels, the effectiveness of the CCP will in part depend on the actual conservation measures taken by producers. In 1995, almost one-fifth of the conservation plans allowed soil erosion levels of more than twice the sustainable level (USDA, 1995c). The 1996 Farm Act allows producers to modify their conservation practices if they can demonstrate that the new practices achieve equal or greater erosion control than the previous measures. Other factors that influence the environmental effectiveness of the CCP include the effort devoted to monitoring and enforcement of compliance,[31] and the number of farmers staying out of farm support programmes because of high compliance costs.

Although the effects of the CCP on farming practices have not yet been assessed, there is evidence that conservation tillage (comprising mulch tillage, ridge tillage and no-tillage) are increasingly being used on erodible soils. Estimates suggest that in 1994 conservation tillage was applied on 43 per cent of the highly erodible land used for the production of corn, cotton, soyabeans and wheat, compared with 27 per cent in 1989 (USDA, 1995a).

The CCP plays an important role in facilitating the transition of land coming out of the Conservation Reserve Program to sustainable agricultural uses. To maintain the environmental gains on CRP land, conservation farming will in many cases have to be adopted. Estimates suggest that three-quarters of the land currently in the CRP would be subject to conservation compliance if returned to production (Osborn, 1995).

Conservation programmes based on "cross-compliance" mechanisms, such as the CCP and the swampbuster and sodbuster provisions, are only effective if the targeted environmental problems are found on farms participating in federal farm support programmes, and only as long as the financial benefits provided by these programmes are large enough to maintain a high level of participation. Until 1996, the potential threat of having to forego the benefits of the deficiency payment programmes provided a strong incentive to comply with conservation programmes. As from 1996, farmers have to comply with the CCP and the swampbuster provision to qualify for market transition payments.

The **Environmental Quality Incentive Program** (EQIP) was established in 1996 to encourage environmental improvements on crop and livestock farms. The pro-gramme is funded at US$130 million in 1996 and at US$200 million annually throughout the 1997-2002 period. Half of the funds are reserved for addressing conservation problems in livestock production. The programme pays farmers up to 75 per cent of the costs of implementing conservation measures, such as improve-ments in manure management, pest management and erosion control. Contracts are for a duration of 5 to 10 years. A competitive offer process is being used to maximise environmental benefits relative to taxpayer costs. EQIP was included in the **Environmental Conservation Acreage Reserve Program**, which serves as an umbrella for the CRP and the WRP, to ensure that the three programmes are operated in a consistent manner (USDA, 1996).

The **Conservation Farm Option** is a pilot programme created by the 1996 Farm Act, which is restricted to producers of wheat, feed grains, cotton and rice receiving production flexibility contract payments. Participants have to implement a conser-vation plan that protects soil, water and related resources, wetlands and wildlife habitat. In return for entering a 10-year contract, farmers receive a single consoli-dated annual payment in lieu of separate CRP, WRP and EQIP payments. Authorised funding of the programme increases from US$7.5 million in 1996 to US$62.5 in 2002, amounting to a total of US$197.5 million.

Other conservation programmes introduced in 1996 include: the **Grazing Lands Conservation Initiative,** a scheme that provides technical and educational assistance to owners of private grazing land and for which US$20 million have been allocated for 1996, US$40 million for 1997 and US$60 million for 1998 and subse-quent years; the **Wildlife Habitat Incentive Program,** which provides US$50 million in cost-sharing assistance during the 1996-2002 period for the improvement of wildlife habitat on the farm; a **flood risk reduction** provision that encourages farmers to idle frequently flooded lands; the **Farmland Protection Program,** which authorises US$35 million in budgetary funds to preserve prime and unique farmland from commercial development; and the **Everglades Agricultural Area,** under which US$300 million are made available for restoration activities in the Everglades, including land acquisition and a swap of Federal land for land in the Everglades ecosystem.

Relationship with agricultural support policies

In principle, agricultural **support policies** may have raised the *cost of the CRP* if they have led to higher cropland rental rates. Also, rental price increases as a result of the supply-reducing effect of CRP itself may have raised the cost of subsequent CRP enrolments. However, it is not clear whether this has actually occurred.

Land withdrawn from wheat production under the CRP has accounted for about 15 per cent of total wheat plantings in recent years. For corn, this proportion has been around 6 per cent, for other feed grains 25 per cent (Australian Bureau of Agricultural and Resource Economics, 1995). In spite of these sizeable figures, the CRP is not an adequate tool for short-term **supply control,** both because of the long duration of the contracts and the "slippage" associated with targeting erodible land. The option to terminate CRP contracts after five years, introduced by the 1996 Farm Act, may have somewhat increased farmers' flexibility to take land out of the scheme in response to world market price developments, but a large-scale return of CRP land to crop production would put the environmental benefits of the programme at risk.

The ARP, on the other hand, proved to be an effective instrument for supply management, even though its effect on cereals plantings depended on farmer participation in commodity programmes. The benefits from commodity programmes were sufficiently high in the first half of the 1990s to ensure that farmers accounting for over 80 per cent of the national base areas of wheat and major feed grains participated in the ARP.

The ARP was indirectly linked to the CRP in terms of its effects on crop production. Decreases in commodity stocks due to increases in CRP acreage translated into reductions in ARP percentages. For this reason, enrolment of wheat base area in the CRP did not reduce the total area planted to wheat. ARP set-aside requirements were also influenced by the *Export Enhancement Program* and the *marketing loan policy* for wheat and feed grains, which has allowed farmers since 1993 to repay price support loans at actual market prices when those prices were below the loan rate.

The CRP may have had negative economic effects on **rural communities** by idling resources in agriculture and the agri-business industry, especially in counties with a high percentage of cropland enrolment. On the other side, in communities where government transfers to farmers under the CRP were in excess of the income lost from retiring land, there may have been a net increase of purchasing power.

Concluding remarks

The **Conservation Reserve Program** was primarily designed to reduce soil erosion on highly erodible land. The criteria for land selection and the requirement to establish a permanent cover on the diverted land reflect this objective. Most of the cropland enrolled in the CRP is covered with forage or native grass. Special conservation practices on CRP land vary substantially across regions, with, for instance, wildlife-improving practices being concentrated in the Great Plains and reforestation in the Southeast. Windbreaks and riparian filter strips cover relatively small areas of high conservation value.

The CRP has significantly reduced soil erosion on the land enrolled in the programme. It may also have contributed to the improvement of wildlife habitat and the protection of water resources. The programme has been modified several times, resulting in tighter conditions for land enrolment, a wider set of environmental objectives, and a greater concentration of diverted land in conservation priority areas. However, there are indications that the benefits of the CRP have been achieved at a relatively high budgetary cost, especially in the early years of the programme, and that the benefit/cost ratio could have been improved through better environmental targeting and a more competitive bidding process.

While land retirement may have been an effective and cost-efficient tool for reducing soil erosion on the most highly erodible land, it is questionable whether a permanent cover was necessary for all land enrolled in the CRP to reduce soil erosion to sustainable levels. There are indications that sufficient erosion control could have been achieved with less restrictive measures, at least on part of the land currently idled under the CRP, by providing incentives for farmers to adopt less erosive cropping patterns, including changes in crop rotation and tillage practices.

Part of the CRP land might eventually be returned to annual cropping. If not carefully managed, this process could undo many of the environmental improvements achieved over the last ten years. While the Conservation Compliance Provision will protect much of the land from erosion when it is returned to production, other environmental benefits, such as wildlife improvements and habitat creation, could be lost.

Although the 1996 Farm Act has increased farmers' flexibility to return CRP land to production in periods of tight grain supplies, the programme remains an inadequate tool for supply management. Apart from the "slippage" associated with retiring highly erodible land, the environmental objectives pursued by the programme require a long-term view, which can conflict with short-term market considerations. Over the medium run, the CRP could possibly achieve an equal or greater amount of environmental benefits on a smaller area, if expiring contracts are replaced by land of higher ecological value and the most effective conservation practices are encouraged on this land.

In the future, the CRP is likely to focus more on environmental issues such as wildlife habitat creation, water quality improvement and wetland protection, in addition to soil erosion. To attain its objectives in an effective and cost-efficient way, the CRP will have to be well integrated with the other conservation measures in agriculture, including those established by the 1996 Farm Act. Many of these measures rely on a whole-farm approach to environmental management in agriculture and are implemented based on farm conservation plans.

The **Acreage Reduction Program** was an effective, short-term supply management tool, yet its environmental benefits were very limited. It did not encourage farmers to target their set-aside obligations to erodible or other environmentally sensitive land. Although land idled under the ARP had to be put in conservation use, the permitted uses included environmentally harmful practices, such as summer fallow. In principle, the same plots of land could be set side for a multi-year period, but year-to-year fluctuations in the set-aside percentage reduced the incentive for investment in conservation measures.

Although there is considerable experience with the management of land diversion programmes in the United States, information about the actual environmental impacts of the programmes is still limited. Apart from estimates of erosion reductions on CRP land, few quantitative assessments have been made of the magnitude of the environmental benefits of the schemes. With the number of agricultural conservation measures increasing, the development of environmental monitoring systems, including farm management and environmental indicators, would seem to be essential for effective programme administration and co-ordination.

NOTES

1. Long- term land diversion in the EU includes set-aside under the forestry scheme and set-aside under agri-environmental regulation (see Chapter 4). Due to the recent implementation of agri-environmental programmes in many EU countries, systematic enrolment figures for set-aside under the agri-environmental regulation were not yet available by mid-1996.

2. The exchange rates used in this document to convert national currency values to US dollars are average annual exchange rates. If the national currency value refers to a particular year, that year's exchange rate is applied; otherwise, 1995 average annual exchange rates are used.

3. Besides the PCP, the *North American Waterfowl Management Plan* (NAWMP) is the second most important initiative in Canada for wildlife preservation in agriculture. The objective of the NAWMP is to protect or restore high-quality wetland habitat in North America. The link between the NAWMP and the PCP enables farmers to receive additional financial assistance for habitat or wetland restoration (Saskatchewan Wetland Conservation Corporation).

4. This estimate is based on a reduction of grain production of 19.5 million bushels per year and includes lower federal premiums for the *Crop Insurance* and *Gross Revenue Insurance Programs,* a reduction in matching contributions to the *Net Income Stabilisation Account,* the proportion of subsidy under *the Western Grain Transportation Act* that is attributable to grains produced on marginal soils now under the PCP, and the reduction in acreage payments under the *Farm Support Adjustment Measures II* due to the difference in payment rates for forage and cropland (PFRA, 1992 and 1993 update).

5. The study was framed in terms of possible modifications to the WGTA as it existed in 1992. The results reported here correspond to the scenario where producers pay full transportation rates and receive no government compensation.

6. GRIP was terminated in Saskatchewan in 1995; in its place, additional government subsidies to the NISA accounts of Saskatchewan producers were made (OECD, 1996).

7. The base area was extended in 1994 to include the linseed area.

8. The set-aside percentage determined each year by the Agriculture Council is the percentage that has to be applied by farmers in their land use decisions. If the Council fails to make decision, the "default" set-aside percentage specified in CR 1765/92

including amendments, comes into effect. Until 1996, the default rates were 15 per cent for rotational set-aside and 20 per cent (18 in the United Kingdom and Denmark) for non-rotational set-aside; in 1996 a uniform rate of 17.5 per cent was adopted.

9. The minimum size restriction facilitates monitoring through remote sensing.

10. In Germany, for instance, the upper limit is set at 33 per cent. (Institut für Europäische Umweltpolitik, 1994.) A special exemption applies to former voluntary five-year set-aside. Producers with land coming out of the five-year scheme may, under certain circumstances, continue to set all of this land aside under the arable aid scheme, even if they exceed the specified set aside ceiling. A lower payment rate applies to the extra set-aside (*additional voluntary set-aside*).

11. The reference yield is determined on the basis of regional yield data for the 1986/87 to 1990/91 period (European Commission, 1993*a*).

12. In England, for instance, a number of larger farmers claimed payments under the simplified scheme in the first year of the arable aid scheme in order to avoid having to set aside land. However, as the payment rates were increased, virtually all large farms joined the main scheme with its set-aside requirement.

13. Independently of this provision, transfers of set-aside obligations can also be allowed if otherwise farmers would not have enough cropland to spread the animal manure produced on the farm. However, this possibility is only offered in Denmark and Germany (European Commission, 1993*b*).

14. For certain crops, soil preparation for the next growing season has to begin before the end of August. If such a crop is to be planted in the year following set-aside, use of fertilisers and certain plant protection products can be permitted on set-aside land after July 15.

15. The figure is for EU-15, excluding Sweden.

16. See also the discussion on input substitution in connection with the United States' land diversion programmes (Chapter 7).

17. Zonal programmes, such as *Environmentally Sensitive Area* schemes, apply to areas which are fairly homogeneous with respect to certain environmental and landscape character-istics. Zonal programmes incorporate normally a combination of several of the meas-ures offered under the agri-environmental regulation.

18. Regarding integrated farming, the whole-farm approach applies only since 1996. Before 1996, participation by farming activity was possible: a farmer could, for example, choose to apply integrated farming methods to crop production and continue as previously with his livestock activity.

19. The *soil loss tolerance level* is the maximum amount of erosion that can be sustained in the long run without a decline in soil productivity.

20. The *erodibility index* measures the inherent erosion potential of the soil in a field.

21. In 1989, for instance, the following lands were eligible for enrolment in the CRP: *i)* cropland with actual erosion in excess of the soil loss tolerance level T *and* potential erosion greater than eight times T, as measured by the erodibility index; *ii)* cropland with actual erosion in excess of three times the soil loss tolerance level; *iii)* all cropland

22. Before, early release of land from the CRP was only possible if it was in the public interest. A breech of contract was subject to a penalty including the repayment of government funds received under the CRP.

23. A *Basic Conservation System* plan is aimed at reducing soil erosion to the soil loss tolerance level. This is a stricter requirement in terms of erosion control than that currently required under the *Conservation Compliance Provision* in order to obtain farm programme benefits on highly erodible land.

24. The 1990 Food, Agriculture, Conservation, and Trade Act linked acreage reduction percentages to the ratio of closing stocks to total (domestic and export) use of US grains.

25. The 1985 Food Security Act specified that at least one-eighth of CRP acreage should be planted to trees.

26. Ribaudo *et al.* analyse the implications for production of land diversions aimed at water quality improvements.

27. In general, land diversion will lead to an increase in the use of non-land inputs only where the (absolute value of the) elasticity of consumer demand for the commodity is smaller than the elasticity of substitution between land and the other inputs. Whether this is the case for US agriculture, is subject to discussion. Floyd (1965), for instance, believed that demand for agricultural raw products is inelastic, whereas the other inputs can, to some extent, be substituted for land. Under such conditions, land set-asides would increase use of non-land inputs on the remaining acreage.

28. For a discussion of the environmental benefits associated with farm forestry, see also OECD (1995b).

29. Only around 16 per cent of the cropland that is vulnerable to groundwater pollution was eligible for the CRP (Ribaudo *et al.*, 1990).

30. The *maximum payment acreage* of a programme crop was calculated as 85 per cent of the crop acreage base less the acreage idled under the ARP. The 15 per cent non-payment acreage was called the "normal flex" acreage (USDA, 1993).

31. By 1992, almost two thousand producers had been found in violation of the conservation compliance and sodbuster provisions, involving some 150 000 acres of land.

The first paragraph (continuation):

in land capability Classes VI, VII or VIII; *iv)* cropland adjacent to water bodies, from 66 to 99 feet wide; *v)* certain cropped wetlands; and *vi)* cropland subject to scour erosion from flooding (Sinner, 1989).

REFERENCES

AGRA EUROPE (1996), Issues of 15 March and 26 April, London.

AGRICULTURE AND AGRI-FOOD CANADA (1995), "The Health of Our Soils. Towards Sustainable Agriculture in Canada", Ottawa.

AGRICULTURE AND AGRI-FOOD CANADA (1993), "Agricultural Policies and Soil Degradation in Western Canada: An Agro-Ecological Economic Assessment: Conceptual Framework", Policy Branch, Technical Report 2/93.

ANSELL, D.J. and S.A. VINCENT (1994), "An Evaluation of Set-Aside Management in the European Union With Special Reference to Denmark, France, Germany and the UK", Centre for Agricultural Strategy Paper No. 30, University of Reading.

AUSTRALIAN BUREAU OF AGRICULTURAL AND RESOURCE ECONOMICS (1995), US Farm Bill 1995: US agricultural policies on the eve of the 1995 farm bill, Canberra.

BALDOCK, D., G. BEAUFOY, G. BENNETT and J. CLARK (1993), Nature Conservation and New Directions in the EC Common Agricultural Policy, Institute for European Environmental Policy, London and Arnhem.

BALDOCK, D., G. COX, P. LOW and M. WINTERS (1990), "Environmentally Sensitive Areas: Incrementalism or Reform?", Journal of Rural Studies, Vol. 6(2), pp. 143-62.

BENBROOK, C.M. (1995), "Impacts of the American Farmland Trust Conservation Reserve Program Recommendations: Preliminary Estimates and Description of a CRP Policy Impacts Simulator", American Farmland Trust, June.

BENBROOK, C.M. (1994), "Environmental Stewardship Through Cross-Compliance: A Review of the US Experience in Agriculture", Report to the Environment Bureau, Agriculture and Agri-Food Policy Branch, Agriculture Canada.

BERMEJO, I. (1994), "Conservación de sistemas adehesados en Extremadura", Agricultura, No. 738, pp. 40-43.

BROUWER, F. and S. VAN BERKUM (1996a), "The Environmental Effects of Land Diversion Schemes: The Case of the European Union", Consultancy paper for the OECD.

BROUWER, F. and S. VAN BERKUM (1996b), "CAP and Environment in the European Union", Paper presented at the Conference on European Agriculture at the Crossroads: Competition and Sustainability, held at the University of Crete, Greece, 9-12 May.

BUNDESMINISTERIUM FÜR ERNÄHRUNG, LANDWIRTSCHAFT UND FORSTEN (1996), "Die Europäische Agrarreform. Pflanzlicher Bereich. Flankierende Massnahmen", Das Bundesministerium für Ernährung, Landwirtschaft und Forsten informiert, No. 421-2/96.

CAREY, M., R. HEIMLICH and R. BRAZEE (1990), "A Permanent Wetland Reserve. Analysis of a New Approach to Wetland Protection", United States Department of Agriculture, Economic Research Service, Agriculture Information Bulletin No. 610.

CLUB DE BRUXELLES (1995), *Agriculture and the Environment in Europe,* Bruxelles.

COMMISSION OF THE EUROPEAN COMMUNITIES (1995a), "On the Purpose and the Methods of Application of Extraordinary Set-Aside", COM(95)122.

COMMISSION OF THE EUROPEAN COMMUNITIES (1995b), *The Agricultural Situation in the European Union. 1994 Report,* Brussels/Luxembourg.

COMMISSION OF THE EUROPEAN COMMUNITIES (1994), "The European Union's Action for the Conservation, Management and Sustainable Development of Its Forests", Progress Report on the Implementation of the UNCED Forest Principles and of Agenda 21, DG VI, FK/UNCED/Report 3.

COMMISSION OF THE EUROPEAN COMMUNITIES (1993a), "Reform of the CAP and Its Implementation", CAP Working Notes 1993, VI/2024/93-EN.

COMMISSION OF THE EUROPEAN COMMUNITIES (1993b), "CAP-Agricultural Information. Set-Aside. A Brief Guide to the Existing Rules", Memo 43/1993, October.

COMMISSION OF THE EUROPEAN COMMUNITIES (1992), *Towards Sustainability: A European Community Programme of Policy and Action in Relation to the Environment and Sustainable Development,* Brussels.

COMMUNAUTÉS EUROPÉENNES (1995), "Description générale des mécanismes du marché commun agricole. Avant-projet de budget général des Communautés européennes pour l'exercice 1996. Fonds européen d'orientation et de garantie agricole, section garantie".

COUNCIL FOR THE PROTECTION OF RURAL ENGLAND (1995), "Set-Aside", London, March.

COUNCIL FOR THE PROTECTION OF RURAL ENGLAND (1994), "Transfers of Set-Aside Obligations", London, June.

DUNN, C.P., F. STEARNS, G.R. GUNTENSPERGEN and D.M. SHARPLE (1993), "Ecological Benefits of the Conservation Reserve Program", *Conservation Biology,* Vol. 7(1), pp. 132-39.

EUROPEAN COMMUNITIES (1992), "Council Regulation (EEC) No. 2078/92 of 30 June 1992 on agricultural production methods compatible with the requirements of the protection of the environment and the maintenance of the countryside", *Official Journal of the European Communities,* No. L 215/85, 30 July 1992.

EUROPEAN COMMUNITIES (1992), "Council Regulation (EEC) No. 2080/92 of 30 June 1992 instituting a Community aid scheme for forestry measures in agriculture", *Official Journal of the European Communities,* No. L 215/96, 30 July 1992.

ENVIRONMENTAL MANAGEMENT ASSOCIATES (1993), "Environmental Assessment of NISA: Final Report", Report prepared for Agriculture Canada, Calgary, Alberta, September.

FLOYD, J.E. (1965), "The Effects of Farm Price Supports on the Returns to Land and Labour in Agriculture", *American Journal of Agricultural Economics,* Vol. 47, pp. 148-58.

GIRTH, J. (1990), "Common Ground. Recommendations for Policy Reform to Integrate Wildlife Habitat, Environmental and Agricultural Objectives on the Farm", Report commissioned by Wildlife Habitat Canada, Environment Canada and Agriculture Canada.

GODDEN, B., P. CLOTUCHE, V. VAN BOOL, M. PENNINCKX and A. PEETERS (1994), "Jachère. Gestion et arrière effet", *Agricontact,* No. 266, November.

GRAY, R. and B. PADDOCK (1993), "Land Set Aside: Is it Food Security?", *Canadian Journal of Agricultural Economics,* Vol. 41, pp. 441-51.

GRAY, R., H. FURTAN, G. CONACHER, R. STEINKE and V. MANALOOR (1993), "Set Aside Options for Western Canada – Final Report", Agricultural Economics University of Saskatchewan, March.

HAWKE, N. and N. KOVALEVA (1994), "Environmental Protection Under the New Set-Aside Scheme", *The ALA Bulletin,* December.

HAWKE, N., A. ROBINSON and N. KOVALEVA (1993), "Set-Aside: Its Legal Framework and Environmental Protection", *Environmental Law and Management,* October, pp. 153-58.

HEIMLICH, R. and J. MELANSON (1995), "Wetlands Lost, Wetlands Gained", *National Wetlands Newsletter,* Vol. 17(3), May-June.

HERTEL, T.W. and M.E. TSIGAS (1991), "General Equilibrium Analysis of Supply Control in US Agriculture", *European Review of Agricultural Economics,* Vol. 18, pp. 167-91.

HODGE, I. (1992), "Supply Control and the Environment: The Case for Separate Policies", *Farm Management,* Vol. 8(2), pp. 65-72.

INSTITUT NATIONAL DE LA RECHERCHE AGRONOMIQUE (INRA) (1995), *L'environnement à l'INRA,* November.

INSTITUT FÜR EUROPÄISCHE UMWELTPOLITIK (1994), *Umweltgerecthe Landwirtschaft. Nachhaltige Wege für Europa,* Economica Verlag, Bonn.

IWAMA, H. and H. OTSUKA (1995), "How do Agricultural Practices Conserve Sloping Lands in Japan?", National Institute of Agri-Environmental Sciences. Paper presented at the OECD Consultation Meeting on Agri-Environmental Indicators, Paris, 9-10 October.

JAPANESE MINISTRY OF AGRICULTURE, FORESTRY AND FISHERIES (1995), "Assessment of Flood Mitigation and Water Conservation Capacities of Farmlands and Forests", Paper presented at the OECD Consultation Meeting on Agri-Environmental Indicators, Paris, 9-10 October.

MCRAE, T., N. HILLARY, R.J. MACGREGOR and C.A.S. SMITH (1995), "Role and Nature of Environmental Indicators in Canadian Agricultural Policy Development", Paper presented to the Symposium of the Resource Policy Consortium on Environmental Indicators, held at the World Bank, Washington D.C., 12 June.

OECD (1997), *Environmental Indicators for Agriculture,* Paris.

OECD (1996), *Agricultural Policies, Markets and Trade in OECD Countries. Monitoring and Evaluation 1996,* Paris.

OECD (1995*a*), *Sustainable Agriculture. Concepts, Issues and Policies in OECD Countries,* Paris.

OECD (1995*b*), *Forestry, Agriculture and the Environment,* Paris.

OFFICE FÉDÉRAL DE L'AGRICULTURE (1995), "Évaluation des matières premières renouvelables", Berne, November.

OFFICE OF TECHNOLOGY ASSESSMENT (1995), *Agriculture, Trade and Environment. Achieving Complementary Policies,* Congress of the United States.

OSBORN, C.T., F. LLACUNA and M. LINSENBIGLER (1992), "The Conservation Reserve Program. Enrolment Statistics for Sign-Up Periods 1-11 and Fiscal Years 1990-92", USDA, Economic Research Service, Statistical Bulletin No. 843, November.

OSBORN, C.T. (1995), "Changes in Store for CRP", *Agricultural Outlook,* September.

PALOMO, J.A. (1994), "Agricultura y Medio Ambiente en Extremadura", *Agricultura,* No. 738, pp. 44-45.

PARRIS, K. and J. MELANIE (1993), "Japan's agriculture and environmental policies: time to change", *Agriculture and Resources Quarterly,* Vol. 5(3), pp. 386-99.

PIMENTEL, D., C. HARVEY, P. RESOSUDARMO, K. SINCLAIR, D. KURZ, M. MCNAIR, S. CRIST, L. SHPRITZ, L. FITTON, R. SAFFOURI and R. BLAIR (1995), "Environmental and Economic Costs of Soil Erosion and Conservation Benefits", *Science,* Vol. 267, pp. 1117-23.

PRAIRIE FARM REHABILITATION ADMINISTRATION (1992), "Potential Impact of Permanent Cover Programs on Federal Government Expenditures", PFRA Policy Analysis Service, July.

DE PUTTER, J. (1995), "The Greening of Europe's Agricultural Policy: The 'Agri-Environmental Regulation' of the MacSharry Reform", Ministry of Agriculture, Nature Management and Fisheries, and Agricultural Economics Research Institute, The Hague, The Netherlands.

RAYMENT, M. (1995), "A Review of the CAP Arable Reforms", The Royal Society for the Protection of the Birds, Arable Policy Paper No. 1.

RIBAUDO, M.O., C.T. OSBORN and K. KONYAR (1994), "Land Retirement as a Tool for Reducing Agricultural Nonpoint Source Pollution", *Land Economics,* Vol. 70(1), pp. 77-87.

RIBAUDO, M.O., D. COLACICCO, L.L. LANGNER, S. PIPER and G.D. SCHAIBLE (1990), "Natural Resources and Users Benefit from the Conservation Reserve Program", USDA, Economic Research Service, Agricultural Economic Report No. 627, January.

RYGNESTAD, H. and R. FRASER (1996), "An Assessment of the Impact of Choice of Set-Aside Scheme on Nitrogen Use", Paper presented at the Agricultural Economics Society Annual Conference, held at the University of Newcastle Upon Tyne, March.

SASKATCHEWAN WETLAND CONSERVATION CORPORATION (no date), "The Permanent Cover Program".

SCHOU, J.S. (1995), "Set-Asides in Denmark", Danish Institute of Agricultural and Fisheries Economics.

SEBILLOTE, M., S. ALLAIN, T. DORÉ and J.M. MEYNARD (1993), "La jachère et ses fonctions agronomiques, économiques et environnementales. Diagnostic actuel", *Courrier de l'environnement de l'INRA*, No. 20, September.

SINNER, J. (1989), "Alternatives for US Soil Conservation Policy: Land Retirement versus Targeted Incentive Programs for Improved Farming Practices", Draft Paper, National Center for Food and Agricultural Policy, Resources for the Future, Washington D.C.

SINNER, J. (1990), "Soil Conservation: We Can Get More For Our Tax Dollars", *Choices,* Second Quarter.

SMITH, R.B.W. (1995), "The Conservation Reserve Program as a Least-Cost Land Retirement Mechanism", *American Journal of Agricultural Economics,* Vol. 77, pp. 93-105.

TERRESTRIAL AND AQUATIC ENVIRONMENTAL MANAGERS LTD. (1992), "An Environmental Assessment of Land Use Changes Due to Proposed Modifications of the Western Grain Transportation Act", Report prepared for Agriculture Canada, Bureau for Environmental Sustainability, Melville, Saskatchewan, December.

TOBEY, J.A. and K.A. REINERT (1991), "The Effects of Domestic Agricultural Policy Reform on Environmental Quality", *Journal of Agricultural Economics Research,* Vol. 43, pp. 20-28.

TYRCHNIEWICZ, A. and A. WILSON (1994), "Sustainable Development for the Great Plains – Policy Analysis", International Institute for Sustainable Development, Winnipeg.

UNITED KINGDOM MINISTRY OF AGRICULTURE, FISHERIES AND FOOD (1996*a*), "Environmental Improvements to Set-Aside Rules", *MAFF News Release,* 26 April.

UNITED KINGDOM MINISTRY OF AGRICULTURE, FISHERIES AND FOOD (1996*b*), "UK to Press Commission on Guaranteed Set-Aside", *MAFF News Release,* 15 January.

UNITED KINGDOM MINISTRY OF AGRICULTURE, FISHERIES AND FOOD (1995), "European Agriculture: The Case for Radical Reform", Working paper prepared for the Minister of Agriculture, Fisheries and Food's CAP Review Group.

USDA (UNITED STATES DEPARTMENT OF AGRICULTURE) (1996), *Agricultural Outlook Special Supplement: Provisions of the 1996 Farm Bill,* April.

USDA (1995*a*), "Updates on Agricultural Resources and Environmental Indicators: Tillage and Cropping Systems on Highly Erodible Land", Economic Research Service, Natural Resources and Environment Division, Update No. 6.

USDA (1995*b*), "Meeting Conservation Goals: What Can be Learned?", *Agricultural Outlook,* April.

USDA (1995*c*), "1995 Farm Bill: Guidance of the Administration", Section 3: Conservation and the Environment, May.

USDA (1995*d*), "Natural Resource Inventory. Graphic Highlights of Natural Resource Trends in the United States between 1982 and 1992", Natural Resource Conservation Service, April.

USDA (1995*e*), "Updates on Agricultural Resources and Environmental Indicators: 1995 Cropland Use", Economic Research Service, Natural Resources and Environment Division, Update No. 12.

USDA (1994a), "Analysis of the Conservation Reserve Program: Farmers' Plans and Environmental Targeting Issues", Economic Research Service, April.

USDA (1994b), "Agricultural Resources and Environmental Indicators", Economic Research Service, December.

USDA (1993), "Agricultural Resources. Cropland, Water and Conservation. Situation and Outlook Report", Economic Research Service, May.

USDA (1992), "Farmers Sign on to Wetlands Program", *Agricultural Outlook,* October.

UNITED STATES GENERAL ACCOUNTING OFFICE (1995a), "Conservation Reserve Program. Alternatives Are Available for Managing Environmentally Sensitive Cropland", Report to the Senate Committee on Agriculture, Nutrition and Forestry, February.

UNITED STATES GENERAL ACCOUNTING OFFICE (1995b), "Agricultural Conservation. Status of Programs That Provide Financial Incentives", Report to the Senate Committee on Agriculture, Nutrition and Forestry, April.

UNITED STATES GENERAL ACCOUNTING OFFICE (1989), "Farm Programs. Conservation Reserve Program Could Be Less Costly and More Effective", Report to the Senate Committee on Agriculture, Nutrition and Forestry, November.

VERCHERAND (1996), "La jachère: une maîtrise de la production coûteuse", *Économie Rurale,* No. 232, March-April.

WAGNER, R. (1995), "Die zukünftige Nutzung ertragsschwacher Standorte in den neuen Bundesländern", Berichte über Landwirtschaft, Vol. 73(3), pp. 466-508.

WESTERN OPINION RESEARCH (1994), "PFRA – Permanent Cover Program. Final Report", Western Opinion Research, Inc., Winnipeg, March.

WILLIAMSON, J. (1993), "CAP Reform Set-Aside: Environmental Friend or Foe?", International Agriculture and Trade Reports – Europe. Situation and Outlook Report, USDA, Economic Research Service, September.

MAIN SALES OUTLETS OF OECD PUBLICATIONS
PRINCIPAUX POINTS DE VENTE DES PUBLICATIONS DE L'OCDE

AUSTRALIA – AUSTRALIE
D.A. Information Services
648 Whitehorse Road, P.O.B 163
Mitcham, Victoria 3132   Tel. (03) 9210.7777
Fax: (03) 9210.7788

AUSTRIA – AUTRICHE
Gerold & Co.
Graben 31
Wien I   Tel. (0222) 533.50.14
Fax: (0222) 512.47.31.29

BELGIUM – BELGIQUE
Jean De Lannoy
Avenue du Roi, Koningslaan 202
B-1060 Bruxelles Tel. (02) 538.51.69/538.08.41
Fax: (02) 538.08.41

CANADA
Renouf Publishing Company Ltd.
5369 Canotek Road
Unit 1
Ottawa, Ont. K1J 9J3   Tel. (613) 745.2665
Fax: (613) 745.7660

Stores:
71 1/2 Sparks Street
Ottawa, Ont. K1P 5R1   Tel. (613) 238.8985
Fax: (613) 238.6041

12 Adelaide Street West
Toronto, QN M5H 1L6   Tel. (416) 363.3171
Fax: (416) 363.5963

Les Éditions La Liberté Inc.
3020 Chemin Sainte-Foy
Sainte-Foy, PQ G1X 3V6 Tel. (418) 658.3763
Fax: (418) 658.3763

Federal Publications Inc.
165 University Avenue, Suite 701
Toronto, ON M5H 3B8   Tel. (416) 860.1611
Fax: (416) 860.1608

Les Publications Fédérales
1185 Université
Montréal, QC H3B 3A7   Tel. (514) 954.1633
Fax: (514) 954.1635

CHINA – CHINE
Book Dept., China Natinal Publications
Import and Export Corporation (CNPIEC)
16 Gongti E. Road, Chaoyang District
Beijing 100020 Tel. (10) 6506-6688 Ext. 8402
(10) 6506-3101

CHINESE TAIPEI – TAIPEI CHINOIS
Good Faith Worldwide Int'l. Co. Ltd.
9th Floor, No. 118, Sec. 2
Chung Hsiao E. Road
Taipei   Tel. (02) 391.7396/391.7397
Fax: (02) 394.9176

**CZECH REPUBLIC –
RÉPUBLIQUE TCHÈQUE**
National Information Centre
NIS – prodejna
Konviktská 5
Praha 1 – 113 57   Tel. (02) 24.23.09.07
Fax: (02) 24.22.94.33
E-mail: nkposp@dec.niz.cz
Internet: http://www.nis.cz

DENMARK – DANEMARK
Munksgaard Book and Subscription Service
35, Nørre Søgade, P.O. Box 2148
DK-1016 København K   Tel. (33) 12.85.70
Fax: (33) 12.93.87

J. H. Schultz Information A/S,
Herstedvang 12,
DK – 2620 Albertslung   Tel. 43 63 23 00
Fax: 43 63 19 69
Internet: s-info@inet.uni-c.dk

EGYPT – ÉGYPTE
The Middle East Observer
41 Sherif Street
Cairo   Tel. (2) 392.6919
Fax: (2) 360.6804

FINLAND – FINLANDE
Akateeminen Kirjakauppa
Keskuskatu 1, P.O. Box 128
00100 Helsinki

Subscription Services/Agence d'abonnements :
P.O. Box 23
00100 Helsinki   Tel. (358) 9.121.4403
Fax: (358) 9.121.4450

***FRANCE**
OECD/OCDE
Mail Orders/Commandes par correspondance :
2, rue André-Pascal
75775 Paris Cedex 16 Tel. 33 (0)1.45.24.82.00
Fax: 33 (0)1.49.10.42.76
Telex: 640048 OCDE
Internet: Compte.PUBSINQ@oecd.org

Orders via Minitel, France only/
Commandes par Minitel, France
exclusivement : 36 15 OCDE

OECD Bookshop/Librairie de l'OCDE :
33, rue Octave-Feuillet
75016 Paris   Tel. 33 (0)1.45.24.81.81
33 (0)1.45.24.81.67

Dawson
B.P. 40
91121 Palaiseau Cedex   Tel. 01.89.10.47.00
Fax: 01.64.54.83.26

Documentation Française
29, quai Voltaire
75007 Paris   Tel. 01.40.15.70.00

Economica
49, rue Héricart
75015 Paris   Tel. 01.45.78.12.92
Fax: 01.45.75.05.67

Gibert Jeune (Droit-Économie)
6, place Saint-Michel
75006 Paris   Tel. 01.43.25.91.19

Librairie du Commerce International
10, avenue d'Iéna
75016 Paris   Tel. 01.40.73.34.60

Librairie Dunod
Université Paris-Dauphine
Place du Maréchal-de-Lattre-de-Tassigny
75016 Paris   Tel. 01.44.05.40.13

Librairie Lavoisier
11, rue Lavoisier
75008 Paris   Tel. 01.42.65.39.95

Librairie des Sciences Politiques
30, rue Saint-Guillaume
75007 Paris   Tel. 01.45.48.36.02

P.U.F.
49, boulevard Saint-Michel
75005 Paris   Tel. 01.43.25.83.40

Librairie de l'Université
12a, rue Nazareth
13100 Aix-en-Provence   Tel. 04.42.26.18.08

Documentation Française
165, rue Garibaldi
69003 Lyon   Tel. 04.78.63.32.23

Librairie Decitre
29, place Bellecour
69002 Lyon   Tel. 04.72.40.54.54

Librairie Sauramps
Le Triangle
34967 Montpellier Cedex 2 Tel. 04.67.58.85.15
Fax: 04.67.58.27.36

A la Sorbonne Actual
23, rue de l'Hôtel-des-Postes
06000 Nice   Tel. 04.93.13.77.75
Fax: 04.93.80.75.69

GERMANY – ALLEMAGNE
OECD Bonn Centre
August-Bebel-Allee 6
D-53175 Bonn   Tel. (0228) 959.120
Fax: (0228) 959.12.17

GREECE – GRÈCE
Librairie Kauffmann
Stadiou 28
10564 Athens   Tel. (01) 32.55.321
Fax: (01) 32.30.320

HONG-KONG
Swindon Book Co. Ltd.
Astoria Bldg. 3F
34 Ashley Road, Tsimshatsui
Kowloon, Hong Kong   Tel. 2376.2062
Fax: 2376.0685

HUNGARY – HONGRIE
Euro Info Service
Margitsziget, Európa Ház
1138 Budapest   Tel. (1) 111.60.61
Fax: (1) 302.50.35
E-mail: euroinfo@mail.matav.hu
Internet: http://www.euroinfo.hu//index.html

ICELAND – ISLANDE
Mál og Menning
Laugavegi 18, Pósthólf 392
121 Reykjavik   Tel. (1) 552.4240
Fax: (1) 562.3523

INDIA – INDE
Oxford Book and Stationery Co.
Scindia House
New Delhi 110001   Tel. (11) 331.5896/5308
Fax: (11) 332.2639
E-mail: oxford.publ@axcess.net.in

17 Park Street
Calcutta 700016   Tel. 240832

INDONESIA – INDONÉSIE
Pdii-Lipi
P.O. Box 4298
Jakarta 12042   Tel. (21) 573.34.67
Fax: (21) 573.34.67

IRELAND – IRLANDE
Government Supplies Agency
Publications Section
4/5 Harcourt Road
Dublin 2   Tel. 661.31.11
Fax: 475.27.60

ISRAEL – ISRAËL
Praedicta
5 Shatner Street
P.O. Box 34030
Jerusalem 91430   Tel. (2) 652.84.90/1/2
Fax: (2) 652.84.93

R.O.Y. International
P.O. Box 13056
Tel Aviv 61130   Tel. (3) 546 1423
Fax: (3) 546 1442
E-mail: royil@netvision.net.il

Palestinian Authority/Middle East:
INDEX Information Services
P.O.B. 19502
Jerusalem   Tel. (2) 627.16.34
Fax: (2) 627.12.19

ITALY – ITALIE
Libreria Commissionaria Sansoni
Via Duca di Calabria, 1/1
50125 Firenze   Tel. (055) 64.54.15
Fax: (055) 64.12.57
E-mail: licosa@ftbcc.it

Via Bartolini 29
20155 Milano   Tel. (02) 36.50.83

Editrice e Libreria Herder
Piazza Montecitorio 120
00186 Roma   Tel. 679.46.28
Fax: 678.47.51

Libreria Hoepli
Via Hoepli 5
20121 Milano　　　　Tel. (02) 86.54.46
　　　　　　　　　　Fax: (02) 805.28.86

Libreria Scientifica
Dott. Lucio de Biasio 'Aeiou'
Via Coronelli, 6
20146 Milano　　　　Tel. (02) 48.95.45.52
　　　　　　　　　　Fax: (02) 48.95.45.48

JAPAN – JAPON
OECD Tokyo Centre
Landic Akasaka Building
2-3-4 Akasaka, Minato-ku
Tokyo 107　　　　　　Tel. (81.3) 3586.2016
　　　　　　　　　　Fax: (81.3) 3584.7929

KOREA – CORÉE
Kyobo Book Centre Co. Ltd.
P.O. Box 1658, Kwang Hwa Moon
Seoul　　　　　　　　Tel. 730.78.91
　　　　　　　　　　Fax: 735.00.30

MALAYSIA – MALAISIE
University of Malaya Bookshop
University of Malaya
P.O. Box 1127, Jalan Pantai Baru
59700 Kuala Lumpur
Malaysia　　　　　　Tel. 756.5000/756.5425
　　　　　　　　　　Fax: 756.3246

MEXICO – MEXIQUE
OECD Mexico Centre
Edificio INFOTEC
Av. San Fernando no. 37
Col. Toriello Guerra
Tlalpan C.P. 14050
Mexico D.F.　　　　　Tel. (525) 528.10.38
　　　　　　　　　　Fax: (525) 606.13.07
E-mail: ocde@rtn.net.mx

NETHERLANDS – PAYS-BAS
SDU Uitgeverij Plantijnstraat
Externe Fondsen
Postbus 20014
2500 EA's-Gravenhage　Tel. (070) 37.89.880
Voor bestellingen:　　Fax: (070) 34.75.778

Subscription Agency/Agence d'abonnements :
SWETS & ZEITLINGER BV
Heereweg 347B
P.O. Box 830
2160 SZ Lisse　　　　Tel. 252.435.111
　　　　　　　　　　Fax: 252.415.888

**NEW ZEALAND –
NOUVELLE-ZÉLANDE**
GPLegislation Services
P.O. Box 12418
Thorndon, Wellington　Tel. (04) 496.5655
　　　　　　　　　　Fax: (04) 496.5698

NORWAY – NORVÈGE
NIC INFO A/S
Ostensjoveien 18
P.O. Box 6512 Etterstad
0606 Oslo　　　　　　Tel. (22) 97.45.00
　　　　　　　　　　Fax: (22) 97.45.45

PAKISTAN
Mirza Book Agency
65 Shahrah Quaid-E-Azam
Lahore 54000　　　　Tel. (42) 735.36.01
　　　　　　　　　　Fax: (42) 576.37.14

PHILIPPINE – PHILIPPINES
International Booksource Center Inc.
Rm 179/920 Cityland 10 Condo Tower 2
HV dela Costa Ext cor Valero St.
Makati Metro Manila　Tel. (632) 817 9676
　　　　　　　　　　Fax: (632) 817 1741

POLAND – POLOGNE
Ars Polona
00-950 Warszawa
Krakowskie Prezdmiescie 7　Tel. (22) 264760
　　　　　　　　　　Fax: (22) 265334

PORTUGAL
Livraria Portugal
Rua do Carmo 70-74
Apart. 2681
1200 Lisboa　　　　　Tel. (01) 347.49.82/5
　　　　　　　　　　Fax: (01) 347.02.64

SINGAPORE – SINGAPOUR
Ashgate Publishing
Asia Pacific Pte. Ltd
Golden Wheel Building, 04-03
41, Kallang Pudding Road
Singapore 349316　　Tel. 741.5166
　　　　　　　　　　Fax: 742.9356

SPAIN – ESPAGNE
Mundi-Prensa Libros S.A.
Castelló 37, Apartado 1223
Madrid 28001　　　　Tel. (91) 431.33.99
　　　　　　　　　　Fax: (91) 575.39.98
E-mail: mundiprensa@tsai.es
Internet: http://www.mundiprensa.es

Mundi-Prensa Barcelona
Consell de Cent No. 391
08009 – Barcelona　　Tel. (93) 488.34.92
　　　　　　　　　　Fax: (93) 487.76.59

Libreria de la Generalitat
Palau Moja
Rambla dels Estudis, 118
08002 – Barcelona
　　(Suscripciones) Tel. (93) 318.80.12
　　(Publicaciones) Tel. (93) 302.67.23
　　　　　　　　　　Fax: (93) 412.18.54

SRI LANKA
Centre for Policy Research
c/o Colombo Agencies Ltd.
No. 300-304, Galle Road
Colombo 3　　　　　Tel. (1) 574240, 573551-2
　　　　　　　　　　Fax: (1) 575394, 510711

SWEDEN – SUÈDE
CE Fritzes AB
S–106 47 Stockholm　Tel. (08) 690.90.90
　　　　　　　　　　Fax: (08) 20.50.21

For electronic publications only/
Publications électroniques seulement
STATISTICS SWEDEN
Informationsservice
S-115 81 Stockholm　Tel. 8 783 5066
　　　　　　　　　　Fax: 8 783 4045

Subscription Agency/Agence d'abonnements :
Wennergren-Williams Info AB
P.O. Box 1305
171 25 Solna　　　　Tel. (08) 705.97.50
　　　　　　　　　　Fax: (08) 27.00.71

Liber distribution
Internatinal organizations
Fagerstagatan 21
S-163 52 Spanga

SWITZERLAND – SUISSE
Maditec S.A. (Books and Periodicals/Livres
et périodiques)
Chemin des Palettes 4
Case postale 266
1020 Renens VD 1　　Tel. (021) 635.08.65
　　　　　　　　　　Fax: (021) 635.07.80

Librairie Payot S.A.
4, place Pépinet
CP 3212
1002 Lausanne　　　　Tel. (021) 320.25.11
　　　　　　　　　　Fax: (021) 320.25.14

Librairie Unilivres
6, rue de Candolle
1205 Genève　　　　　Tel. (022) 320.26.23
　　　　　　　　　　Fax: (022) 329.73.18

Subscription Agency/Agence d'abonnements :
Dynapresse Marketing S.A.
38, avenue Vibert
1227 Carouge　　　　Tel. (022) 308.08.70
　　　　　　　　　　Fax: (022) 308.07.99

See also – Voir aussi :
OECD Bonn Centre
August-Bebel-Allee 6
D-53175 Bonn (Germany) Tel. (0228) 959.120
　　　　　　　　　　Fax: (0228) 959.12.17

THAILAND – THAÏLANDE
Suksit Siam Co. Ltd.
113, 115 Fuang Nakhon Rd.
Opp. Wat Rajbopith
Bangkok 10200　　　　Tel. (662) 225.9531/2
　　　　　　　　　　Fax: (662) 222.5188

**TRINIDAD & TOBAGO, CARIBBEAN
TRINITÉ-ET-TOBAGO, CARAÏBES**
Systematics Studies Limited
9 Watts Street
Curepe
Trinidad & Tobago, W.I. Tel. (1809) 645.3475
　　　　　　　　　　Fax: (1809) 662.5654
E-mail: tobe@trinidad.net

TUNISIA – TUNISIE
Grande Librairie Spécialisée
Fendri Ali
Avenue Haffouz Imm El-Intilaka
Bloc B 1 Sfax 3000　Tel. (216-4) 296 855
　　　　　　　　　　Fax: (216-4) 298.270

TURKEY – TURQUIE
Kültür Yayinlari Is-Türk Ltd.
Atatürk Bulvari No. 191/Kat 13
06684 Kavaklidere/Ankara
　　　　　　　　Tel. (312) 428.11.40 Ext. 2458
　　　　　　　　　　Fax : (312) 417.24.90

Dolmabahce Cad. No. 29
Besiktas/Istanbul　　Tel. (212) 260 7188

UNITED KINGDOM – ROYAUME-UNI
The Stationery Office Ltd.
Postal orders only:
P.O. Box 276, London SW8 5DT
Gen. enquiries　　　　Tel. (171) 873 0011
　　　　　　　　　　Fax: (171) 873 8463

The Stationery Office Ltd.
Postal orders only:
49 High Holborn, London WC1V 6HB
Branches at: Belfast, Birmingham, Bristol,
Edinburgh, Manchester

UNITED STATES – ÉTATS-UNIS
OECD Washington Center
2001 L Street N.W., Suite 650
Washington, D.C. 20036-4922
　　　　　　　　　　Tel. (202) 785.6323
　　　　　　　　　　Fax: (202) 785.0350
Internet: washcont@oecd.org

Subscriptions to OECD periodicals may also
be placed through main subscription agencies.

Les abonnements aux publications périodiques
de l'OCDE peuvent être souscrits auprès des
principales agences d'abonnement.

Orders and inquiries from countries where Dis-
tributors have not yet been appointed should be
sent to: OECD Publications, 2, rue André-Pascal, 75775 Paris Cedex 16, France.

Les commandes provenant de pays où l'OCDE
n'a pas encore désigné de distributeur peuvent
être adressées aux Éditions de l'OCDE, 2, rue
André-Pascal, 75775 Paris Cedex 16, France.

12-1996

OECD PUBLICATIONS, 2, rue André-Pascal, 75775 PARIS CEDEX 16
PRINTED IN FRANCE
(51 97 01 1) ISBN 92-64-15366-7 – No. 49233 1997